Cost Planning of Buildings

Seventh Edition

Douglas J.
PhD, FRICS

Peter S. Brandon
DSc, MSc, FRICS, ASAQS

Jonathan D. Ferry
BSc (Hons)

Blackwell
Science

Blackwell Science Ltd, a Blackwell Publishing company
Editorial Offices:
Blackwell Science Ltd, 9600 Garsington Road, Oxford OX4 2DQ, UK
Tel: +44 (0) 1865 776868
Blackwell Publishing Inc., 350 Main Street, Malden, MA 02148-5020, USA
Tel: +1 781 388 8250
Blackwell Science Asia Pty Ltd, 550 Swanston Street, Carlton, Victoria 3053, Australia
Tel: +61 (0)3 8359 1011

First published in Great Britain by Crosby Lockwood & Sons Ltd 1964
Second edition (metric) published 1970
Third edition published 1972
Fourth edition published by Granada Publishing
Fifth edition published 1984
Sixth edition published by BSP Professional Books 1991
Seventh edition published by Blackwell Science Ltd 1999

5   2006

ISBN-10: 0-632-04251-6
ISBN-13: 978-0-632-04251-7

Library of Congress Cataloging-in-Publication Data
Ferry, Douglas J. (Douglas John)
  Cost planning of buildings/Douglas J. Ferry, Peter S. Brandon, and Jonathan D. Ferry.—7th ed.
     p.cm.
  Includes bibliographical references and index.
  ISBN 0-632-04251-6 (pbk.)
  1. Building—Estimates.   I.   Brandon, P.S. (Peter S.) II. Ferry, Jonathan D. III. Title.
TH435.F36 1999   99-19451
692—dc21   CIP

A catalogue record for this title is available from the British Library

Set in 10/12pt Times
by DP Photosetting, Aylesury, Bucks
Printed and bound in India
by Replika Press Pvt. Ltd.

The publisher's policy is to use permanent paper from mills that operate a sustainable forestry policy, and which has been manufactured from pulp processed using acid-free and elementary chlorine-free practices. Furthermore, the publisher ensures that the text paper and cover board used have met acceptable environmental accreditation standards.

For further information on
Blackwell Publishing, visit our website:
www.blackwellpublishing.com

Contents

Preface

Over the last 15–20 years the bulk of the industry's clientele has moved from the public sector (in which formal cost planning techniques first developed) to the private sector. At the same time there has been a growth in refurbishment and rebuilding work, compared with new build.

In these circumstances, market conditions and budgeting now have a greater influence on the teaching of the subject, and sophisticated cost modelling a somewhat lesser place. This is reflected in the arrangement of this new edition, which has been re-organised so that techniques of cost planning are dealt with in the context of the tasks to be undertaken at each stage of the process:

> **Stage 1 – The Brief and the Budget** deals with the establishment of the budget from the standpoint of the client's needs and resources.
> **Stage 2 – Designing to the Budget** deals with the design of the project within the client's budget.
> **Stage 3 – Controlling the Cost** deals with maintaining the budget through the production of working drawings and the construction of the project.

At each stage theoretical examples are given, for the first time, with real-life examples kindly contributed by Dearle and Henderson Consulting, in which the names and addresses have of course been altered to preserve anonymity.

It might be useful here to mention the American term 'value engineering', which is increasingly being used by the UK construction industry. This can be seen as a subset of cost planning. As practised in the UK, it usually concentrates on Stage 3 of the process outlined above, with an emphasis on cost cutting.

Finally, the book seeks to provide an essential groundwork in all aspects of the discipline, while containing ample material as a basis for seminar work, discussion and class development. It cannot give full coverage to each of the topics dealt with in principle, and a reader who wishes to explore any of these in greater depth will need to refer to specialist works. A suggested list for further reading and references is given at the end of most chapters. However, the most useful sources of reading are the weekly magazines, especially *Building* and *The Architects' Journal*, as these are always up to date.

Douglas J. Ferry
Peter S. Brandon
Jonathan D. Ferry

Introduction

Chapter 1
Introduction

Need for cost estimating

From the earliest times people have needed some idea of what a new building was going to cost before they started work on it. The New English Bible says:

> 'Would any of you think of building a tower without first sitting down and calculating the cost, to see whether he could afford to finish it? Otherwise, if he has laid its foundations and then is not able to complete it, all the onlookers will laugh at him. "There is the man" they will say "who started to build and could not finish".'
>
> (St Luke, Ch 14)

Forecasting the cost of a building, however, is not the same thing as planning the cost, any more than a weather forecaster on television can be said to be planning the weather – in both cases things may turn out very differently from what was expected for reasons outside the forecaster's control. Nevertheless, until about 200 years ago rough-and-ready forecasting was all that was needed, because:

- Profitability was not an issue.
- Most major building work was undertaken either as an act of religious faith or by the very rich for their own pleasure.
- The building process itself comprised a series of independent and sequential craft activities of which the costs had become established and known over a long period of time.

Even so miscalculations occurred – the building of Blenheim Palace almost bankrupted the Duke of Marlborough, and it was not at all unknown for prospective owners of buildings to suffer the fate of the man in the Bible.

With the advent of the Industrial Revolution in the late eighteenth century, however, something better was needed for a number of reasons:

- The traditional settled economic and social order was turning into something more sophisticated and dynamic.
- The projects themselves were of an increasing technological complexity and innovation.

3

- The people commissioning large building projects were increasingly cost conscious, being either industrialists concerned with profitability, government bodies concerned with accountability, or joint stock companies concerned with both.

Price-in-advance system

In order to deal with the needs of this new situation the price-in-advance system was developed, which in a modified form is still in widespread use today:

- Responsibility for the execution of the whole project was handed over to a 'general contractor' at a previously agreed price.
- The building owner and the architect got rid of many of the problems of organising the construction, and secured a firm cost commitment before starting work.
- But complete drawings and specification of the work had to be prepared before prices could be sought.

Prices were usually obtained by competitive tender; the documents would be sent to a number of general contractors who would submit sealed offers to be opened by the architect at a time and date set in advance. The job would normally be given to the firm whose tender was the lowest.

In order for contractors to prepare their tenders they had to calculate the quantities of labour and material which would be needed to build the project. This was done by measuring the work off the drawings, and since these figures should be the same for all the firms tendering (unless a mistake were made) it was obviously more economic for this task to be undertaken by one person on behalf of them all.

Origin of the quantity surveyor

This was the origin of the quantity surveyor (QS), whose task was the preparation of a 'bill of quantities' (BQ). The BQ set out the quantities of finished work in the project, from which each builder would calculate the hours of labour and quantities of raw material (allowing for waste) that were thought to be needed. The QS's fee was paid by the successful tenderer.

A difficulty immediately arose. The success of the tendering system depended upon there being true competition, and architects were concerned that if the tenderers had to meet in order to agree the appointment of a QS they might also use the meeting to agree who should get the job. The chosen tenderer could then submit an inflated price which would include a pay-off for all the others (who would have submitted even more inflated prices to ensure that the chosen tenderer won). Therefore architects started sending out a BQ with the drawings and specifications, so depriving the tenderers of an excuse to meet together, although the successful builder was still responsible for paying the QS who had been appointed by the architect. From this point, quantity surveying developed in the UK (and in many of its overseas dependencies) into a fully recognised profession.

Cost estimating prior to tender

Firm price competitive tendering gave the prospective building owner a reliable budget, and an opportunity to withdraw before any actual expenditure on building was undertaken. However, a good deal of other expenditure would have been incurred by this stage which could not be recovered if the project was unable to go ahead because tenders were too high:

- If the project were abandoned the architect and quantity surveyor would already have completed most of their work and would require payment for this. (The only people who would have worked without charge would have been the builders, who would not expect to be able to recover their tendering costs – a sore point with many of them.)
- The client would probably have bought the land for the project and this money would now be locked up in an unproductive asset.
- The client might have made plans for the use of the building which would have to be abandoned with consequent loss.

An alternative to abandonment, of course, would be complete redesign on a more modest scale and the seeking of fresh tenders but this would involve a further loss of time and would not relieve the client of the abortive expenditure on the initial scheme. Clients therefore needed to have a good idea of what the building was likely to cost before design had progressed very far, and for this reason an 'approximate estimate' would be prepared. This had to be done at an early stage, when even the basic drawings would exist in outline only and there would be little or no supporting documentation.

Although the usual terms of appointment required the architect to prepare this estimate, it was usually done by the QS on the architect's behalf, as the QS had a greater expertise and a wider range of available cost data. It was not a lengthy task and was normally done free of charge as a favour to the architect, who had usually been responsible for nominating the QS to the client.

Single price rate estimating

The cost-estimating techniques used by the QS were basically 'single price rate' estimates; the size of the building was calculated in square or cubic measure and the result multiplied by a single price rate to give the estimated cost. These traditional methods will be dealt with later in this book, since they are still valid today in some circumstances as part of a comprehensive system of cost planning and control.

However, such an estimate suffered from a major defect; during the development of the design from sketch plan to working drawings it was not possible to relate the cost of the work which was shown on the detailed drawings to the estimate. The QS, while preparing the BQs, might feel that the working drawings were showing something rather grander and more extravagant than had been envisaged when the estimate was prepared. However, because of the lack of any detail in the estimate it was not possible to prove this to the architect or client except by preparing and pricing 'approximate quantities'. These were a simplified version of a formal BQ; major items were grouped and measured together, while less important items were ignored and dealt with by loading the prices of major items.

The disadvantages of such a course are obvious:

- The task required extra time and work and would involve an additional fee.
- The drawings had to be fairly complete and the whole of the building had to be dealt with.
- Even if the approximate quantities when prepared showed a far higher total than the original estimate, the reasons could not be easily traced. It still could not be seen whether any particular part of the building was extravagantly designed, and there was nothing to show where the cost could be cut most economically.
- In addition, even if it were possible to identify the problem areas the architect would obviously resist an expensive major redesign at this late stage.

As a rule, therefore, approximate quantities would not be prepared but the job sent out to tender by the architect, who would just hope for the best. If the tender proved to be too high the price would have to be cut wherever it could be done easily without having to alter too many drawings, usually by cutting the more high-class parts of the specification and substituting lower quality. This was not at all the same thing as producing an economically efficient design.

Advent of cost planning

Nevertheless the system worked reasonably well until the years following the Second World War, when the art of accurate single price rate estimating became increasingly difficult to practise because of:

- unsettled economic conditions;
- the use of non-traditional designs;
- the increasing proportion of the cost represented by engineering services such as electrical and heating installations.

Public authorities in particular, who were at this time responsible for most building work, demanded value and efficiency in building design, and a system of cost planning and control during design stage came into vogue. This system was called 'elemental cost planning' and enabled the cost of the scheme to be monitored during design development, as well as facilitating the rational allocation of costs amongst the various elements of the building structure, finishes and services. The technique is still used for these purposes today, but more importantly it formed the basis upon which a whole range of cost planning practices was developed.

Unfortunately, no sooner had success been achieved in rationalising the economic approach to the obtaining of tenders than the price-in-advance system itself began to deteriorate, for two reasons:

- First the British system of tendering upon sophisticated BQs enabled tenders to be obtained on a basis which was theoretically that of a completed and worked-out design, but in practice could sometimes be little more than a hypothetical construct. Design development, with an enormous number of changes both great and small, would therefore be undertaken during the construction stage. The contractor would find it difficult to organise the work properly but had ample excuse to charge a price which differed substantially from the original offer.

- Second the changed social, economic and moral climate made it very difficult for a general contractor to quote a firm price for a major project in competition which would adequately allow for all the risks involved. Rampant inflation, industrial action, social legislation and financial irresponsibility showed clearly that society was very different from that which had fostered the Victorian concept of price-in-advance.

In these circumstances alternative methods of building procurement came into favour. In many of these there is no place for formal BQs and no such concept as a 'tender price', but quantity surveyors are still usually appointed on such projects to act as cost controllers, because of their expertise on costs and their manipulation. However, the importance of cost planning and control is such that members of the architectural, engineering and building professions need to have a working knowledge of the principles, and some of them may come to specialise in this aspect of their work and see themselves as cost planners, as do some of their colleagues on the continent of Europe. The development of integrated computer systems which manipulate cost as one of a number of design parameters tends to assist this process, as does the undertaking of work elsewhere in the European Community and overseas.

Two further factors in developing more flexible cost planning techniques in the UK have been:

- The increased proportion of refurbishment and rebuilding projects and a smaller proportion of new 'green field' work – for which the original systems were designed.
- The general shift from a public to a private sector clientele, with less emphasis on the accountability of a construction programme and more emphasis on the profitability of individual projects.

In these circumstances cost planning has now developed into a total process, which:

- starts at the point where a client first thinks of commissioning a building
- continues until the building is completed and paid for

whatever method of procurement is being used.

Summing-up

From the earliest times people have needed some idea of what a new building was going to cost before they started work on it. Until about 200 years ago rough-and-ready forecasting was all that was needed, because profitability was not an issue and the costs of building had become established and known over a long period of time.

During the eighteenth century a system of building contractors tendering in competition came into being. This was the origin of the quantity surveyor (QS), whose task was the preparation of a 'bill of quantities' (BQ). The BQ sets out the quantities of finished work in the project, from which each tenderer can calculate the hours of labour and quantities of raw material (allowing for waste) that are thought to be needed.

The system of single price rate preliminary estimating prior to tendering still worked reasonably well until the years following the Second World War, when it was upset by:

- unsettled economic conditions
- the use of non-traditional designs, and
- the cost of services' installations

and a system of cost planning and control during design stage came into vogue. This system is called 'elemental cost planning' and enables the cost of a scheme to be monitored during design development. The development of integrated computer systems which manipulate cost as one of a number of design parameters tends to assist this process.

Further reading

Cooke (1996) *Economics and Construction*. Macmillan, London.

Gruneberg (1997) *Construction Economics*. Macmillan, London.

Nisbet, J. (1961) *Estimating and Cost Control* (especially Chapter 1 Background). Batsford, London.

Raftery, J. (1991) *Principles of Building Economics*. Blackwell Science, Oxford.

Seeley, I.H. (1996) *Building Economics*, 4th edn. Macmillan, London.

Thompson, F.M.L. (1968) *Chartered Surveyors: The Growth of a Profession*, pp. 64–93. Routledge & Kegan Paul, London.

Chapter 2
The Impact of Information Technology

The advent of information technology (IT) has had a considerable impact on all areas of life within recent years, and this holds equally true for the field of cost planning, as well as the office environment in which the present-day cost planner will be working. This chapter is not intended to provide a definitive analysis of the development of information technology in general, or list all the applications of IT that cost planners are likely to come into contact with and utilise in the general course of their professional work. The subject is constantly being updated, and in any case is too wide to warrant such an undertaking. The purpose of this chapter is therefore to describe some key developments that exemplify the manner in which information technology has developed, and is continuing to develop, to support the subject area.

It will be suggested in later chapters that computer technology is likely to continue to play a large part in improving cost modelling techniques. The power of the machine to hold vast quantities of information and recall and manipulate that information must be of interest to all those charged with providing cost advice. Certainly the techniques described in detail later in the book would be difficult to implement or would be uneconomical in practice unless computers were employed to undertake the calculations.

It should be noted, however, that IT has a wider remit than just the application of computer modelling. The SERC IT applications panel for the Environment Committee defined IT research as the application and development of technology which enables improved data capture, transmission, manipulation or interpretation of information in support of decision-makers or the manufacture of an improved product. If this definition is accepted then there are a large number of systems on the market which have arisen from IT research and which are now supporting commercial activity within the field of building economics and financial management. Almost without exception the supporting technology is computer based, although strictly speaking IT encompasses a wider range of applications. The widespread everyday use of plain-paper facsimiles, modems, and e-mail has resulted in 'instantaneous' communication that has never previously been possible or expected. It could be argued that this has resulted in a compression of timescales and deadlines that are not necessarily always to the advantage of those charged with providing professional advice.

Whilst the widespread use of computers has undoubtedly brought benefits in laborious and mundane tasks that previously would have taken some considerable time to be completed manually, this has to some extent been offset by the accelerated expected timescales referred to above. However, before we look in more detail at some of the ranges of applications to the cost planner, it is worth considering the difference between humans and computers. These differences are summarised in Table 1.

Table 1 gives a good indication of what machines should not be called upon to do with the current state of the technology. Computers are good at recall and calculation but have difficulty with creative processes and handling information which is unstructured. Humans, on the other hand, tend to be slow at calculation and less reliable on recall. The objective must be to harness the strengths of each in order to get the best system. It should also be an evolutionary process so that consultants and practitioners accept the technology as it develops, can trust the output and feel in control.

For this reason the majority of computer systems have adopted manual approaches as a starting point in their development. Most manual processes provide a framework or reference point to which, or from which, consultants can add their expertise. They provide structure and form but do not dictate a solution. Elemental cost planning, for example, uses a standard classification system and historic unit rates but the consultant can vary specifications, quantities and rates as is felt appropriate to the new situation. The wider economics issues can be envisaged, as well as the nature of the client, the features and location of the site, etc., which it would be difficult to encompass within a machine.

The computer has no knowledge of the world, other than the very limited amount given by the programmer or user, and has no means of self-reference so it does not 'understand' why it is doing things. These issues will be important as software develops and the computer appears to be replicating human decision-making. It is always wise to keep in mind the limitations of the technology to avoid the creation of an oppressive tool.

At a more mundane level computers are not very economic where a large amount of detailed information needs to be input before a solution can be obtained. Measurement for bills of quantities (BQs) is an example of such a problem. The process is similar to any estimating process as follows:

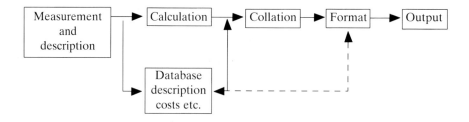

The problem is that of inputting the many thousands of dimensions and descriptions. Apart from measuring there is the problem of checking them and many would argue that a manual method such as 'cut and shuffle' can be just as quick and cheaper. This problem is likely to be lessened in the future as the

Table 1   Summary of comparison between humans and computer.

Characteristics	Machine	Human
Speed of operation	Extremely fast and superior to human in most respects	Slower at physical and simple mental work
Stamina	Can sustain operation *ad infinitum* and limited only by reliability of mechanical and electrical devices	Unable to sustain long periods of work without rest
Accuracy and consistency	Extremely accurate, consistent and reliable. Poor at error correction	Unreliable and inconsistent particularly when dealing with repetitive work. Good at error correction
Senses	Consistent sensitivity. Finds difficulty in sensing from a variety of sources simultaneously. Wider range of sensitivity, dependent upon sensing device	Ability to combine sensory powers. Senses can be affected by environment, drugs etc.
Memory	Generally smaller but very accurate storage and recall	Large memory but subject to memory loss and error in recall
Overload	Sudden breakdown	Gradual degradation
Cognitive ('knowing') processes	Follows instructions with complete accuracy, performing logical and/or arithmetic operations	May follow instructions precisely or in a haphazard fashion. May misunderstand them
	Logic largely a matter of determining whether a statement is true or false. Therefore mainly a process of comparison with given information	Processes may be interrupted by 'creative leaps' short-circuiting a tedious procedure
	Man made criteria required for judging whether true or false	Perceiving, remembering, imagining, conceiving, judging, reasoning affected by feelings, emotions, motivation and so on
Logic	Good at arithmetic operations but can only draw and compare simple analogies	Good at comparing and judging unlike things, at inductive logic and at drawing analogies/metaphors directly relevant to the problem in hand
		Good at making complex decisions

Contd

Table 1 Contd

Characteristics	Machine	Human
Input	Machine input exceptionally fast but human input slow	Slow at character reading but fast at sensory recognition
Output	Very fast, neat and tidy. Output to backing store exceptionally fast. Lack of flexibility in manipulation	Slow output at all times either by speech or hand. Very flexible in manipulation

increasing use of scanning, digitising and direct CAD links becomes more widespread. It is only when the same information is used and manipulated for other tasks that the real benefit of the machine can be seen. In the case of priced BQs the same information can be used for cost analyses, valuations, final accounts, financial reports and cashflows. If repetition of input can be avoided at each stage then substantial savings can be made. This is true not only within a firm itself, but also within the construction industry as a whole.

History of IT concepts related to cost planning

It has already been stated that early use of computer power was to speed up manual processes. These manual processes were developed to compensate for human inadequacies such as slow links, poor memory and poor speed of calculation. The machine could undertake computation at enormous speed and this was seen as a major advantage. Unfortunately very little analysis of the manual processes was undertaken. If it had been investigated it would have been found that calculation and collation represented a fairly small proportion of the total process. Much time is spent in discussion, in searching through literature and diagnosing drawings and other information. The benefits are limited therefore to greater efficiency in perhaps only 15% of the total time. It was only where the application was essentially all calculation that real return could be seen. In the UK the turning point was probably the NEDO Price Adjustment Formula in the early 1980s. Some firms and local authorities were able to claim a payback period of less than a year, on both hardware and software, just on this one operation.

Once firms had machines in their possession their computer literacy improved and they began to look for other ways of harnessing the machines' potential. Prior to 1980 there had been a continuous period of development of BQ computer systems for over two decades. At one time over 35 different systems were being used or developed. However it was not until the microcomputer arrived that a mass market was discovered. The reduction in hardware and software cost plus a shortage of labour to undertake routine work encouraged many firms to re-evaluate use of computers. Software packages were still adopting manual concepts but as time went on the computer was used to integrate information and act as a data source.

By the mid-1980s a large number of firms had purchased computers and were using them for early cost forecasting. The BCIS introduced its *Online* service for access to its data bank. Final accounts, estimating, valuations and other routine tasks were now undertaken commonly by computer. Computer-aided draughting was also gaining in popularity and 3-D modelling of buildings also became more commonplace in larger practices.

Towards the end of the 1980s 'ELSIE', the first commercial expert system related to building economics and construction management, became available and is now quite widely used. 'ELSIE' has since been redefined as a suite of 'Lead Consultant Systems', which are now marketed in the UK by Engineering Technology of Shardlow Hall, London Road, Shardlow, Derbyshire DE72 2GP. Developed by the RICS in collaboration with the University of Salford it pointed the way to the future potential of machines to provide consultancy advice in conjunction with humans, and is described in more detail later in this chapter. Ostensibly it follows traditional manual practices but at the same time is far more efficient, consistent and reliable.

Research into IT

In the future we can expect a continued greater level of integration of software packages across disciplines. One requirement will be a common way of describing buildings in the machine so that all consultants can access a common data model for their requirements. It would appear at the moment that an 'object oriented' approach offers the best solution, but only time will tell.

Research into cost planning techniques over the last two decades has largely centred on or relied upon computer techniques. In the beginning researchers transferred manual techniques to the machine. Subsequently they saw the potential for sensitivity analysis, repeating the calculations tirelessly, changing one or more variables at a time to see the effect. Techniques such as regression and simulation began to be developed, but few of them were adopted in practice on a large scale. A major problem was that they were 'data hungry' and therefore required a high overhead in maintaining their reliability. More importantly they were 'black box' in nature, i.e. information was put into the machine and information was output but the process in between was hidden from the user. Consequently users were being asked to accept that the computer model was perfect (which it wasn't), and at the same time they were barred from bringing their own knowledge and expertise to bear on the problem. Very few practitioners were, or are, willing to place their faith in such models when their clients' and their own livelihood depend on the result. Acceptability is a key factor in modelling which does not appear to have been given a high priority among researchers.

During the late 1980s and 1990s emphasis swung towards the potential of knowledge based systems and in turn the potential for linking with conventional computer models to see whether 'intelligence' can be brought into these systems to relieve mundane work and improve consistency and performance.

The new millennium will no doubt see continued major development in IT, which will further change the techniques employed, the nature of the professions and client expectations. It should be expected that greater standardisation and greater integration of systems will continue to erode the traditional boundaries between disciplines. The power of the computer and integrated software may mean that one person or firm can control a wider sphere of activity beyond their normal discipline. This will signal a race between firms to control and manage the technology. It will be interesting to see which professions maintain their status and which firms survive.

Four key applications

In order to illustrate this wide subject four key applications impacting on building economics are described.

The BCIS Online *database*

It will be seen later, in Chapter 13, how the RICS Building Cost Information Service has led the way in providing useful and pertinent cost information to practitioners in the UK. This service originated as a mailing of hard copy information on economic indicators, indices and labour rates as well as the circulation of cost analyses supplied by subscribing members. It was a natural progression from paper format to supplying the information on a screen at the end of a telephone line and since 1984 subscribers to the BCIS *Online* service have been able to obtain most of the information described in Chapter 13 from their own personal computer with a modem. Thus the time and space requirements of a paper-based system can be reduced. Since new data are being fed into the system all the time the *Online* users have access to the latest information within the database.

At the time of writing a major extension of the system is being planned; the information currently included is listed in Table 2.

The importance of having access to the latest information is most obvious with the indices, especially during times when the market is uncertain. Analyses are added to the system daily and *Online* users will have the latest examples at their fingertips. As well as being able to display and print the latest information, subscribers to BCIS *Online* can update analyses to show cost at current or projected prices. Data can also be exported for use in other applications.

Perhaps as important as fast access to data is the potential for an operator to use the computing power of the machine to undertake manipulation of the data. Hence it is possible to call up one or more analyses similar to a new project and go through a fairly conventional cost planning routine. The real advance, however, is being able to fine tune an estimate quickly and efficiently. Several packages are available which assist the cost planner in this task, such as:

CATOPro Cost Planning from Elstree Computing
Datamaster from Masterbill
Estimate from Construction Software

Table 2 Information currently available in the BCIS *Online* database.

Cost analyses	Elemental
	Group elemental
	Total building level
Indices	Tender price indices
	Regional tender prices
	Cost indices
	Output price indices
	Retail price indices
	Maintenance and cleaning cost indices
	Energy cost indices
	Trade cost indices
	House rebuilding cost index
Briefing and economic indicators such as construction industry output and new orders	
Average building prices	$£/m^2$
	Average element prices
	Average functional unit
Tender price studies	Location factor
	Contract value
	Building type factors
Average percentage of preliminary and other contract conditions	
Daywork rates and build-ups	
Wage agreements	

as well as the Approximate Estimating Package from BCIS. Each package will have its own special features, but they all provide the basic functions of accepting details of the new project, choosing cost analyses to be manipulated (whether from the cost planner's own sources or from BCIS), updating and adjusting to provide an elemental estimate and cost plan which can be amended subsequently.

The system is now well established and other software houses are also developing packages that can access BCIS information and manipulate data in different ways. There is little doubt that databases of cost and their integrated programs will become a major feature of the cost consultant's life in years to come.

Computer aided design

This topic is of course extremely wide ranging, although it is often thought of in terms of draughting packages only. However it covers all those aspects of a building which can be modelled in the computer and which can assist the design team in their decision-making process. It can, therefore, include energy evaluation, structural calculations and cost evaluation.

Generally accepted CAD packages, such as AutoCAD, model the building in three dimensions with solid geometry and can undertake a number of different evaluations. A series of 'layers' is used for different elements of the building, such as walls, or services installations, which can be built up layer upon layer to give a complete representation of the building. Alternatively layers can be stripped away so that only the area of interest is displayed.

The three dimensional systems allow more information to be held with regard to the graphic information. For example walls will not be defined in plan as parallel lines, but rather as solids having height as well as width and depth. The user can then choose to look at an elevation, which merely requires the computer to work out the way the wall solids will appear from another angle. Perspectives can be produced in the same way. Sophisticated 3-D 'walk-throughs' and 'fly-bys' can similarly be undertaken to take a prospective client on a 'tour' of the building, before any construction has commenced. If further information is attached to the solid, such as a description or identification code, or a U-value or cost then other evaluations can take place. It is relatively easy to schedule like items, or to calculate the thermal efficiency of the external envelope, or cost the items that are listed (if of course the costs are proportional to quantity). In addition checks can be included in the system to ensure that such things as lines join up properly, the details conform to good practice, that building regulations are complied with and fittings are not forgotten. This extra discipline placed on the designer will in turn encourage the design to be complete. As each entry into the computer is retained in the memory, then short of a power failure, the item will be included in any schedule or evaluation without error. If it is on the 'drawings' it will be on the 'schedule'. Obviously the more the Standard Method of measurement is simplified and moves towards a 'counting' rather than analogue measurement process then the easier it will be to move to comprehensive automatic measurement.

If cost planning (i.e. forecasting and control) is seen as an aid to design then CAD packages are of considerable interest. In essence CAD systems are large databases of information with evaluative and manipulative routines attached. The database contains information on the coordinates of the building geometry, specification of items, location of components, spatial data, performance measures (e.g. U-values) and so forth. If these are available then it follows that quantitative information can be abstracted, costs can be attached to specification items and similar items can be collated and scheduled. Indeed there are several systems that allow this kind of useful information to be produced and therefore replicate the mechanical processes of measurement, collation, computation and pricing adopted when preparing a cost forecast. It is easier to derive the data required for early cost forecasts than for later estimates such as a BQ. The detail of the design model at outline proposals and scheme design stages is still fairly coarse and the complexity of interaction between components is ignored. It is also easier for the machine to 'recognise' boundaries to spaces and walls at this early stage. Nevertheless there are some systems which claim that they can schedule up to 80% of the items for a full BQ – although not necessarily complying with one of the standard methods of measurements. Where the building is formed from a standard kit of parts, arranged in different ways, then it is possible to produce a full BQ with standard price rates attached. One example of this is the link between the McDonnell Douglas 'General Drafting System' (GDS) and ABS Oldacres BQ-Micro package applied to Mobil Oil petrol filling stations. Once design of a standard station is complete the machine will automatically schedule the component items with their quantities and generate a complete tendering document. Additionally, some firms of housebuilders have achieved a great deal of success with such integrated systems, where a small number of design types are constantly repeated. However, the number of components is severely limited, the specification standardised and the designs conform to a 'house style'. It is a much more difficult problem to take a non-standard high-tech laboratory, hospital or office block and generate a full BQ.

CAD systems from the cost planner's point of view

In general the power of CAD from the cost planner's point of view is:

- All members of the design team are working off the same model and there is therefore consistency across all disciplines and everyone should know the current state of the design.
- Computation of quantities is largely automatic.
- Specifications can be attached to quantities and all similar items collated and scheduled.
- Costs can be attached to specifications/quantities to produce broad cost estimates.
- The cost estimate can be altered automatically with changes in design.

This sounds very attractive but there are problems:

- Even with three-dimensional models it is not always easy to differentiate and classify items for scheduling. The technical problems will be overcome in time (thus posing a threat to all measurers), but it will require greater investment and possibly less attachment to standard methods of measurement.

- In cost estimating there is an element of judgement and this tempers the mechanical application of standard unit rates. This moderation process is derived from the expertise of the consultant and should not be underestimated. It would require much more intelligence in the system to cope with this.

- Scheduling systems do not easily cope with incomplete design. The computer can be used to check whether the design is complete and consistent but it is unusual even with CAD for designs to be complete at tender stage. This could result in omissions or failure of the scheduling system. Humans usually account for these problems by calling on their resource of expertise.

- The time taken to input information can be extensive and the maintenance of the database of standard information, e.g. on costs, is not a trivial exercise. Only when those data are used a number of times for different tasks or where it is likely that there is a high degree of modification will the overhead cost seem worthwhile.

However the advantages can outweigh the problems and particularly at the early stages of design. If several outline solutions can be generated by the designer and quite comparative cost estimates produced for each then the quality of decision making should be improved. Unfortunately most CAD systems were designed as draughting tools at working drawing stage. It was recognised that most of the architect's time was being spent on this time consuming task, whereas the key decisions at the commencement of the design process had comparatively fewer hours allocated. Perhaps the ideal CAD tool would be one in which a simple model was created at the start, which could be manipulated and tested before moving to the next stage of refinement. As the design is refined through the three-dimensional model the machine would be evaluating the building online. Ideally the concept of the 'design cockpit' would be introduced which would provide 'windows' to view the various 'instant' evaluations as design developed. This concept is not remote and various aspects of such a system have already been developed, but not within one package. As standardisation of communication takes place and components are available on standard databases so it will become possible to integrate the current developments.

One requirement for increasing the speed of evolution of these packages will be that the industry accepts a standard description of the components of a building, probably in terms of 'objects' which have certain attributes. This will enable all disciplines to build the models they require based on a known set of parameters. There continues to be much interest in this concept at the present time and it is likely that the research required to develop the idea will be furthered in the next few years. Inevitably it will take time to design and develop the standard objects and associated databases, but this can be done in phases.

Intelligent knowledge based systems (IKBS)

Whereas CAD has been available to the construction industry for the past two decades or more, knowledge based systems or expert systems have only recently been made available commercially.

The British Computer Society specialist interest group in expert systems has offered the following definition:

> 'An expert system is regarded as the embodiment within a computer of a knowledge based component from an expert skill in such a form that the system can offer intelligent advice or take an intelligent decision about a processing function. A desirable characteristic, which many would consider fundamental, is the capability of the system, on demand, to justify its own line of reasoning in a manner directly intelligible to the enquirer.'

There are of course grave dangers in applying the term 'intelligent' to such systems at the present time because they only have certain of the characteristics of simple intelligence and do not begin to compare with that possessed by humans.

The key features of expert systems are the 'knowledge base' and what is usually called the 'inference engine'. The knowledge base is a collection of facts, beliefs and heuristic rules (i.e. rules of thumb) expressed in some form of logical relationship in order to make sense of the individual items. The inference engine undertakes the task of reaching conclusions through the logic. If the knowledge base is sufficiently comprehensive then the system will mimic at least part of what is perceived to be the decision-making process of a human being. Other parts that are now being seen as just as important are an explanation facility and a capacity to refine the knowledge quickly. In time, natural language interfacing with the system, so that users are not so restricted in the way they address the system, will also be considered essential. The technology is at a very early stage. Even though it was developed in the 1960s it was not until the middle 1980s that the supporting tools became commercially available to develop the systems relatively quickly.

Lead Consultant Systems

One of the first commercial expert systems available for the construction industry, worldwide, was developed by the Royal Institution of Chartered Surveyors (Quantity Surveying Division) in collaboration with the University of Salford, Department of Surveying. The systems now called Lead Consultant Systems, but formerly known as ELSIE (L–C for lead consultant), provide consultancy advice at the strategic planning stage of the development process, i.e. before design takes place. They take the kind of information available in a client's brief and translate this information into a series of reports. These reports cover the following and are structured in the form of modules within the system:

Initial Budget:	How much will the development cost?
Time:	How long will the development take?
Procurement:	What is the contractual relationship between the parties to the development process?
Appraisal:	What is the profitability of the scheme?

The four modules are linked together (see Fig. 2.1) and common information is transferred from one to the other through a database. At the end of a consultation between 20 and 30 pages of printed report are given which can be handed direct to the client if the operator so wishes. At the present time three broad categories of building type, together with a system for Mechanical and Electrical Engineering Services, are covered by the consultation, but others are being developed.

Because the system uses the knowledge of experts it operates in a similar way to an expert including the models which are adopted. It has been noted before, that cost models, for example, are merely reference points or skeletons to which consultants add their own expertise. The expert system, therefore, uses the same

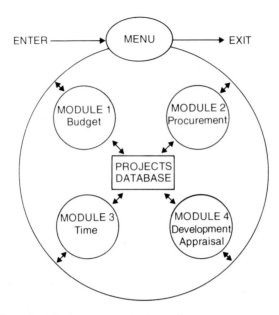

Fig. 2.1 Lead Consultant System: conceptual overview.

approach. It adds to the cost model (in the case of a Lead Consultant System a form of coarse approximate quantities) the extra dimension of the consultant's knowledge. At this early stage consultants have to call on their experience of previous projects and put into the model what they consider to be appropriate and reasonable assumptions. These assumptions are derived by diagnosing the needs of the client from the brief or other method of communication. In reality much of this 'understanding' of the client's brief would come from conversations, visual comprehension of a location and even body language. The unstructured information that comes through the consultant's senses 'triggers' thought patterns, which allow a solution to be arrived at. At the present time, machines do not have the capability to sense and diagnose in the same way as humans. Consequently there will always be a need for consultants to check and possibly modify the solution suggested by a machine until the technology improves very substantially.

In the Lead Consultant initial budget system a solution is postulated by the machine from over 2000 'rules' and the user can then modify the answer by changing up to 150 variables which have been derived. To arrive at the first solution the machine asks between 24 and 30 questions depending on the answers given. It uses rules, logic and inference to generate variables of size, shape and specification and will instantly give a response (in terms of cost) of a change in any one of them.

The relationships between variables are often shown in a paper model called an inference net first so they can be checked by the knowledge provider before they are placed in the machine. An example of an inference net for external walls is shown in Fig 2.2. Note that the final result to the right of the page is a

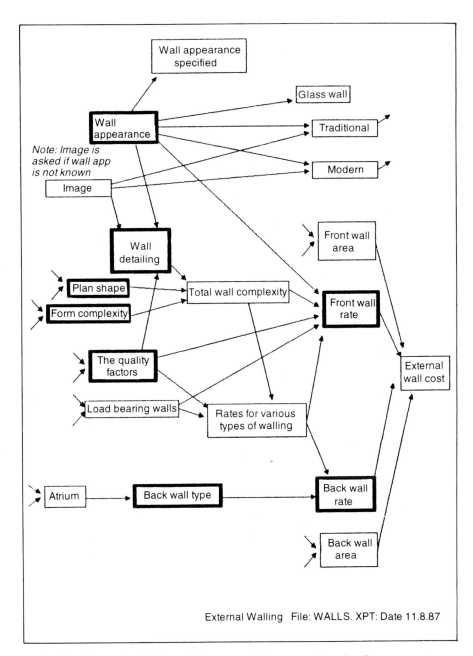

Fig. 2.2 Lead Consultant System: inference net diagram – external walls.

cost based on quantity multiplied by rate. All the other variables are there to determine what these two values should be in order to derive the cost. In the system many of these variables are derived from each other and not requested from the operator. This explains why 150 variables can be determined from asking around only 24–30 questions. The Lead Consultant initial budget module had around 30 of these inference nets to plan and record the knowledge in the system.

Because the knowledge and relationships reside in the machine any changes to other variables are automatically altered should this be required and the machine is always consistent. In use on multi-million pound projects the results have proved to be within plus or minus 5% of the expected tender figure when operated by a knowledgeable consultant. This is rather better than could be expected if several estimators were asked to undertake the same task with the same information. In addition, on more than one occasion the machine has discovered flaws in estimates prepared by skilled cost planners. Additionally, the speed with which Lead Consultant enables reliable initial figures to be derived at a time when information may be scarce is far quicker than could be accomplished manually with the same information.

The advantages for cost modelling of this method are of major importance. Most of the research work established in recent years has resulted in 'black box' techniques. Regression and simulation use statistical relationships, with comparatively few variables to derive a solution. Information is input into the model (the black box) and information comes out at the other end without any intervention or enhancement by the consultant. Nor is the consultant able to view or check, without considerable effort, what is happening. The assumption must be, therefore, that the model is perfect – which we know it isn't. Consequently acceptability of these such models has been low. The transparency and explanation facility of an expert system overcomes these problems to some extent and acceptability is higher. This explains the fast take-up of the Lead Consultant systems which now runs into many hundreds of systems sold.

Construction management systems

The developments in IT have made it more difficult for a subject such as building economics to see itself as a self-contained isolated discipline. It has always been dependent on the actions and views of others and its own knowledge and information is, in turn, used by other consultants and clients for their work. In the context of a building project the figures of the investment analyst or the design from the architect are used as parameters and reference points for the cost planner. The estimates produced from the information provided are then used for assessing profitability levels or modifying the design. It is a quirk of history and the limitations of human ability that have resulted in the diversification of professions into specialisms each with perceived boundaries to the part of the problem which they address.

However, much of the information that each profession uses is common. We have seen already that CAD could provide a common database which all could access to avoid misunderstandings, incorrect measurement and bad communication. In cost planning it now seems rather outdated to consider merely the traditional view of the subject, in terms of estimating and control during design, when this is part of the total financial management of the project. This management covers the whole of the financial processes from the decision to invest to occupation and the choice of demolition or refurbishment. To isolate any particular part of the information flow is inefficient and possibly counter-productive. If every piece of information gathered in the process can be collected and made easily available in a database for use whenever it is required then a major advance has taken place. Instead of measuring the same item, say floor area, several times (by investor, architect, surveyor, contractor, sub-contractor, etc.) the information can be provided instantly for all to use.

The refinement process of information can also be enhanced by using coarse measurements and breaking them down into constituent parts as more knowledge becomes available. It may be possible to do some of this work automatically using knowledge based systems. For example, it is possible now to take an early estimate, refine it as working drawings became available, prepare tender documents and then allocate items to a network of activities to generate a cash flow using standard rates. The RIPAC system developed by Rider Hunts in Adelaide and being sold in the UK by CSSP of Market House, 12–13 Market Square, Bromley, Kent BC1 1NA is an example of a very extensive construction management system which uses the power of a relational database to undertake this work. At the moment much manual input and classification of items is required depending on what output is needed. In time it is possible to foresee some of these processes being automated. After all, the human uses rules to undertake the task and it should be possible to capture at least some of these in the machine.

One of the biggest selling systems in the UK is the CATO (Computer Aided Taking Off) system sold by Elstree Computing. This system uses a relational database and covers a comprehensive range of financial management activities related to a project. Figure 2.3 shows the packages which are integrated to allow estimating, cost monitoring, valuations and final accounts to be brought together.

As the computer technology develops so it will be possible to integrate and automate more of these processes. This will create its own problems. The larger the system the more complex will be the relationships. This cannot be approached in the 'ad-hoc' manner which has often been the pattern of the past. There is a need to design the database with its inter-relationships in a similar way to a manufacturing process or a building. A 'blue print' of the intended systems will provide a framework that would allow the development of databases across the industry in such a way that their data would be compatible. It is important to distinguish the 'design' from the 'physical' database. In a pilot study undertaken at the University of Salford for the RICS Quantity Surveyors Division, a technique called entity-relationship modelling was used to develop a high level model of some of the important relationships between data. Once the model was developed it was then possible to implement it in a physical database using packages such as DBase IV or Oracle without problems of communication. The technique provides a robust method of describing the relationships without ambiguity. This enables any firm to develop a system for its own requirements efficiently and effectively knowing that it will be possible to link with external systems in the future.

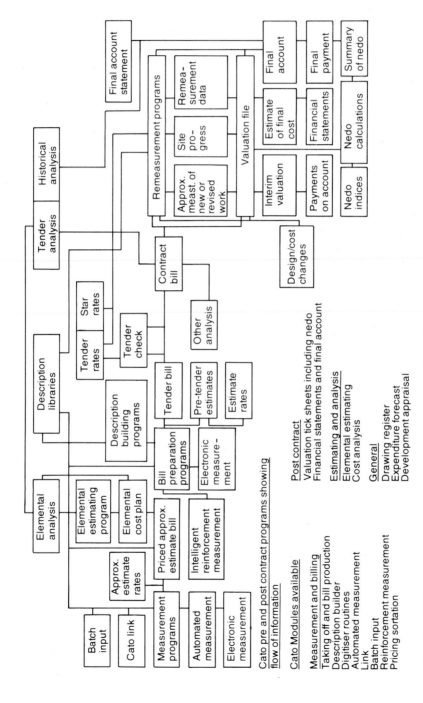

Fig. 2.3   CATO: pre- and post-contract programs showing flow of information.

Summing-up

At a basic level, the widespread use of common computer software, such as spreadsheets, has relieved much of the laborious and mundane nature of data manipulation in everyday situations, for example the preparation and presentation of valuations. In a more general context, since financial management is largely the collection, manipulation and analysis of data, it follows that information technology has much to offer the discipline. Whether it be CAD, knowledge based systems or integrated databases, each provides a major opportunity for advancement. As time goes on so these systems will be linked to other external systems and the information revolution will be well on its way. In a book of this nature it is impossible to do justice to the subject and in some areas the technology is still in its infancy – we do not know how far it can develop. Scientists are working on neural networks and molecular computers. It is sometimes difficult to believe that just over one working lifetime has passed since the first electronic computer. The development of the next 50 years will revolutionise what we do now in all aspects of our life and the subject of this book will be no exception.

Chapter 3
A Three-Stage Cost Planning Strategy

Scope of cost planning and control

A programme of cost planning and control should comprise:

Stage 1: The setting of a budget.
Stage 2: The cost planning and control of the design process.
Stage 3: The cost control of the procurement and construction stages.

Stage 1: The setting of a budget

Systematic cost planning of buildings was first introduced at a time when the national building programme was largely in the public sector. Budgeting was undertaken on the basis of government formulae for the cost of houses, schools, hospitals, etc., and the skill of the cost planner lay in optimising these in relation to the particular project. Today however budgeting is concerned much more with value for money and real-world financing and business decisions. It is altogether a more client-oriented process and its importance in the cost planning process has grown accordingly. It must be emphasised that the purpose of cost planning is not to obtain minimum standards and a 'cheap job', but to budget correctly for a desired standard and then ensure that the resulting budget is spent effectively.

Stage 2: The cost planning and control of the design process

This comprises:

- establishment of the brief;
- investigation of a satisfactory solution;
- cost control of the development of the design.

It is unfortunately true that the amount of time and effort spent by the cost planner on each of these three aspects is often in inverse proportion to their relative importance. Very often effort is almost entirely concentrated on the third aspect, with perhaps a little advice being given on the second. Sometimes the excessive effort required to keep the design development within cost limits actually stems from unwise decisions made at the earlier stages without the benefit of proper cost investigation. In fact, if the strategy is right the design development will usually present few problems. The correct strategy can be summed-up in one phrase: 'zeroing in'. At the early stages the estimates are saying that on past experience a building with certain accommodation and of a certain specification level ought to be able to be built for a certain amount of money, and it is up to the project team to achieve this.

As the design develops the estimates are increasingly saying that the designers are getting it right, so this building as designed can be built for a certain sum.

Estimates given at an early stage carry a fair degree of risk, and as further information becomes available they are almost certain to become subject to amendment. What is important is that whenever this happens the decisions to amend the cost, to amend the scheme, even to abandon the scheme altogether, can be taken without a lot of abortive expenditure having been incurred. Therefore the estimates themselves should not be prepared in any more detail than is relevant to the current stage of progress of the design.

Although the establishment of the brief and the investigation of a satisfactory solution are shown as two separate and consecutive functions, there should be a certain amount of iteration. Design investigation may suggest modifications to the brief which in turn will need to be investigated. This is all to the good, and will probably result in improved performance. Even if it involves, as it will, a good deal of abortive cost-planning work, the cost planner's work at this stage is relatively cheap compared to the potential benefits. The cost control of design development, on the other hand, demands considerable resources, and should not be carried out until a satisfactory solution has been defined and agreed. Substantial iteration between investigation of a solution and cost control of the development of design brings nothing but disadvantages.

In formulating the brief the cost planner should be co-operating, preferably on a team basis, with

- those professionals concerned with the actual building design;
- representatives of the client's organisation or the client's project managers;
- valuation surveyors, accountants, and possibly planners.

As the investigation of a satisfactory design proceeds further into the realms of building configuration the cost planner will become increasingly involved with the designer and the design consultants, including perhaps a construction planner, to the gradual exclusion of the other parties.

As has been stated, the basic principle to be adopted is one of moving from the 'ball-park' estimating of outline proposals to the detailed costing of production drawings in a series of steps. At each of these steps the process must be monitored. Previous assumptions must be checked in the light of further design development, and any necessary modifications made to the estimates or to the design before proceeding to the next stage. As the brief is developed in more detail, and as the design itself develops, it may become apparent that the building which the client wants, with a proper balance between economy and quality, will cost either more or (in theory at any rate) less than has been allowed. It should be possible to reduce or increase the quality of the specification to get back to the original figure, but the cost planner should not automatically assume that the client will want this to be done. The client organisation may be more concerned with getting the building they want than with 5% on the cost; on the other hand they may not, but the decision is theirs. The difference between this approach, and the situation (without cost planning) of a large and unexpected gap between estimate and tender is that clients make their choice consciously with a knowledge of the amount of money which their decision is costing (or saving) them. In this way the time and expense involved in abortive detailed design work can be avoided.

Stage 3: Cost control of the procurement and construction stages

The original development of cost planning, as we have already noted, took place largely within the local authority and central government sectors, where because of the system of approvals in force at that time the amount of the tender was the crucial factor and there was less emphasis on the final cost. Today's clients are not satisfied with this fiction, and the stage of moving from a tender to the actual cost is one which requires at least as much attention as the earlier stages.

The role of the cost planner

Many years ago a quantity surveyor was often not appointed until the architect had prepared the working drawings of the final scheme, or, if appointed earlier, would play no part in things until this stage was reached and work could start on the bills of quantities (BQs). Things have changed since those days, and now the quantity surveyor, or other cost planner, is sometimes appointed before any other professional adviser and takes over total responsibility for the client's financial interests in the scheme. However, most appointments fall between these two extremes.

When the cost planner is appointed there are a number of questions which will have a bearing on the budget, and which need to be answered:

- Who was responsible for the appointment? If the cost planner has been appointed on the recommendation of the client's architect, relations with the client will tend to be conducted through that architect. However, if the appointment has been made by the client as a result of previous experience or outside recommendation the relationship will usually be more direct. Apart from any other factor, a client who decides to appoint a cost planner directly will probably have an above average interest in costs, which will therefore play a predominant part in the scheme design.

- Is the cost planner to be concerned with the total budgeting of the project or limited to particular areas, such as capital expenditure, or building and furnishing costs, or merely net building costs, or expected to do nothing more than merely give an estimate and then prepare a BQ or other contract documentation?

- Who else has been appointed? If other appointments have not been made, the cost planner could assist the cost-oriented client in setting up the team of advisers, and this would obviously be of assistance in the context of total cost control.

- What decisions have already been made, or steps taken? Any decisions which have already been taken, or implemented, will obviously constrain any cost optimisation programme which the cost planner may produce. The cost planner should be very careful about questioning any of these decisions unless advice is definitely being sought or unless they are so fundamentally wrong as to make it impossible to carry out the task for which the cost planner has been appointed. It would only be in the most extraordinary circumstances, and where the cost planner was in a very strong position, that doubts should be cast on specialist professional advice which the client had already received from valuers, accountants, lawyers, etc.

- Is cost control to continue until the completion of the project? In the past it was not unusual for the cost planner's cost control role to finish at the point where a contract was signed with a contractor. In view of the many major problems which can occur subsequently this was a very short-sighted policy and is generally no longer the case.

It will therefore be seen that the extent to which the cost planner will be able to use the various techniques described in this book depends not only upon the priorities of the client but also upon the terms and circumstances of appointment. A cost planner who has been entrusted with the overall cost management of a project must resist the temptation to make firm proposals to the client based upon his or her own elementary knowledge of property values, investment, taxation, etc. Cost planners are not experts in these fields (although a large quantity surveying firm may well employ such experts) and the purpose of education in cost planning courses in this area is to make cost planners aware of circumstances in which these matters may be important, and to help them in briefing specialists and assessing specialist advice.

Office organisation

With regard to the amount of work involved in cost planning, any sensible person would want to keep this to a minimum. The quantity surveyor will soon learn from experience which items are of cost significance, where short cuts can be taken, and where cost checks can be avoided. Experience will also show that the use of computers, standard forms and procedures speeds things up as long as the cost planning staff are familiar with them. For this reason, and also because of the acceleration in working brought about by familiarity with cost and prices, it may be found worthwhile in a large office to concentrate cost planning in a separate department rather than leaving it to be done by the 'takers off' on each project.

In coming to a decision on this point a powerful argument in favour of cost planning by the takers off concerned must be considered: that it will give them familiarity with the project and thus make the actual preparation of the BQ much easier. This proposition, however, implies a staff of quantity surveyors all of whom are fully experienced and practised in both cost planning and BQ preparation techniques. Whether it is as valid as it sounds is therefore open to question; certainly a number of public authorities who originally used this method subsequently changed over to separate departments. There is scope for experiment here, but if the cost planner and the taker off are two different people there must obviously be a close and friendly liaison between them.

Summing-up

Elemental cost planning should ensure that the tender amount is close to the first estimate, or alternatively that any likely difference between the two is anticipated and is acceptable. Elemental cost planning should ensure that the money available for the project is allocated consciously and economically to the various components and finishes. Elemental cost planning does not mean minimum standards and a 'cheap job'; it aims to achieve good value at the desired level of expenditure. Elemental cost planning always involves the measurement and pricing of approximate quantities at some stage of the process.

Stage 1
The Brief and the Budget

Chapter 4
Developers' Motivations and Needs

Developers and development

The carrying out of building work is often called 'development' and the person or organisation responsible for initiating it is the 'developer'. More often in the building industry this person or organisation is called the 'client' or 'building owner', or in some parts of the world the 'proprietor'. In popular use the term 'developer' is often restricted to those who build for profit, but this is not correct. Those who build for their own use or for social purposes are equally developers.

All development arises from a consumer demand. This may be either

- a direct demand: for housing, for offices, for a town hall or library, etc., or
- an indirect demand, that is a demand for something which will require building development in order to satisfy it (such as a demand for TV sets which will need a factory to produce them).

But the demand must always be an economic one. There must be someone willing to foot the bill or who can be induced to do so.

Developers, whether individuals or corporate public or private bodies, will be building either

- for profit, when their approach to cost will be governed by the way in which they expect to receive their reward, usually by leasing or by sale, or
- for use, when their approach will be governed by the actual units of accommodation which they require.

Profit development, social development and user development

In considering cost targets we may distinguish between:

- *profit development*, where it is intended that receipts from the disposal or use of the buildings will more than cover costs,
- *social development*, which is usually in the publicly funded sector,
- *user development*, which covers such projects as a private house, or an office building for an insurance company's own occupation.

There is also *mixed development*, incorporating buildings of more than one of the above types. Because the calculations involved in budgeting for all these types of development are similar it is easy to forget that the basic situations relating to profit development on the one hand, and social or user development on the other, are fundamentally different.

Cost targets for profit development

The setting of cost targets for profit development (e.g. office blocks for letting, housing for letting or sale) is related to free-enterprise economics, and even where there are grants, subsidies, taxation reliefs and so on to be taken into account, the criteria are simple. Even if some non-quantifiable element, like preserving the environment, has to be taken into account it will have to be assessed against its effect on profit. A calculation will soon show how much the developer can afford to spend under the heading of 'building costs' after other items of expenditure (see Chapter 8) such as land costs, professional fees, etc., have been taken into account. There is some flexibility because the standard of building will partly determine the rent or selling price that can be asked, but this is often dictated by the environment and by the cost of the land. It would obviously not be economic to erect a high-cost luxury building for sale or rent in an unpopular neighbourhood, nor to try a low-cost, low-income development in a fashionable district where land costs are high and there is a good income potential.

There is no room for 'hard luck stories' in carrying out profit development, but on the other hand there is the challenge of working in a real situation where time means money. For example:

- early completion of a retail complex may involve substantial extra profits;
- late completion (e.g. after Christmas instead of before Christmas) could be financially disastrous.

Cost targets for social or public sector user development

In social or user development, however, the main object of the exercise is the actual provision of the buildings. It becomes very difficult to set realistic cost targets since there is no definite limit at which an individual building ceases to be possible. In public sector building in particular it is usually possible (though not always politically expedient) to raise whatever sum of money is required for the purpose. Any constraint is usually at a higher level than the individual project, for example, the proportion of the national budget which is allocated to education may determine the amount to be spent on school building as a whole.

Therefore, in order to determine a reasonable cost for this type of building it is necessary to set artificial limits, based upon the cost of similar buildings erected elsewhere. The gross floor area is too crude as a basis for the purposes of this comparison and so various targets based upon user requirements have been established, often by the ministries responsible so as to ensure a nation-wide standard. In the case of schools the unit of cost was the number of 'cost places' (a fictitious number of pupils calculated from the teaching space), while for hospitals the number of beds was once the basic yardstick.

These cost targets were usually determined by a set of artificial standards which had to be adhered to. Although tight in some ways, these usually made exceptions for technical difficulties associated with a particular project. But here we are not in the world of simple profit economics; the early completion of a school or library would usually be a financial burden to the authorities, however welcome it might be to the community.

The cost planning of social development projects, therefore, may resemble the playing of a board-game such as Monopoly, where the architect and cost planners try to win according to a set of rules which have little validity in the real world outside. Sometimes the rules may be very crude indeed, for example:

- A rule that money cannot be transferred from one fund to another (so that it is useless trying to save money on furnishings or running costs by spending extra on the building).
- A rule that money cannot be spent outside the financial year in which it is allocated.
- A rule that no major contract can be let except by competitive tendering on a firm bill of quantities (BQ) with no allowance for cost fluctuations.
- A rule that any financial considerations other than the contract amount are irrelevant.

In these circumstances the skill in cost planning may consist of 'loophole' designing, to take advantage of the regulations in the same way that a clever accountant takes advantage of the tax laws. An example of this, from outside the UK, occurred when a national system for cost control of flat building at one time gave a greater cost allowance for balconies than the actual cost of providing them. The blocks of flats built during this period can be identified by their lavish provision of unnecessary balconies.

It is recognised by government that unrealistic cost rules lead to bad design, and many far-sighted efforts have been made to do away with the cruder kinds of inconsistency:

- by allowing money saved in one direction to be spent in another;
- by trying to bring the assessment of running costs into the cost comparisons;
- by working out very complex cost criteria for such buildings as hospitals instead of 'so much per bed'.

These are commendable attempts to get nearer to reality, but while they have undoubtedly led to some improvement they also tend to make the 'game' more complicated, and the loophole finding more of a challenge to the experts. It becomes increasingly difficult to avoid the 'balcony' type of inconsistency mentioned above.

The consequence of all this is that everybody can become so preoccupied with trying to meet tight cost targets by clever application of the rules, that they lose sight of the social purpose of the whole exercise. The attempts over the years to get as many houses, flats, schools and hospitals as possible out of the budget without the consumer checks on satisfactory standards (such as sales or economic rents) which exist in the private sector have played a large part in creating the massive maintenance programmes with which many public authorities are now faced.

Cost targets for private user development

This third type of development covers such projects as a private house, or an office building for an insurance company's own occupation. User development incorporates some features of both profit and social development. Because of this it is most important to find out what the client's real cost priorities are and to get them defined.

Cost targets for mixed development

This type of development is a mixture of profit and social development, with perhaps some user development. The cost of some of the buildings, or perhaps a proportion of the total cost, would be met by social funding and the rest is intended to make a profit. A common example of this type is a town-centre development incorporating shops and public amenities. The same remark about clients' priorities applies as in the previous case.

Cost-benefit analysis (CBA)

In order to overcome some of the difficulties previously outlined, and to justify the spending of money on public projects, the techniques of cost-benefit analysis were developed. Such analyses attempted to quantify all factors including the various social benefits and disadvantages, and were widely used in connection with traffic and airport schemes and with hospital building. Cost-benefit analysis might be applied, for instance, to a proposal to carry out works to remove a sharp curve and speed restriction from an inter-city railway. The cost of the work could be set against the saving in fuel, and against the wear and tear on equipment caused by braking and re-acceleration. However, it is possible that on these grounds alone the project might not quite be worthwhile. On the other hand, if the curve is removed it might save two minutes each on a million passenger journeys a year – 30,000 man-hours are worth something, and if they are priced and included among the benefits the scheme might now be justifiable in relation to the national economy.

This is all very well, but the railway company is going to incur the costs but is not going to receive the financial benefit of the 30,000 man-hours, so organisations in their position cannot be expected to take this sort of exercise seriously. In addition to this there are the problems of attaching money values to things which cannot be quantified. As an illustration of this, what is the value of a human life? One approach would be to work this out on the basis of the financial contribution which the individual person is expected to make to the economic life of the community, so that a surgeon might be worth hundreds of thousands of pounds while an unemployed labourer might have a negative value. Even on practical grounds this right-wing approach is obviously unacceptable; if you attempted a cost-benefit analysis of a geriatric hospital on this basis you would find that it would be cheaper to let people die and build a mortuary instead! Or, as another less extreme example, you could justify a ring road which saved a few minutes for 'important' people while wasting the time of humble pedestrians.

It was therefore customary to take a notional figure representing the worth of an 'average' person. There was little wrong with this except that such figures, being notional, were conjured out of thin air and could be used in practice to 'prove' that a politically desired result was the right one. As an example, suppose a traffic improvement scheme was going to cost £100,000 a year and was estimated to reduce road accident deaths by three per year. If you cost a life at £50,000 the scheme would obviously be worthwhile; if you cost a life at £20,000 it wouldn't be.

This is not just a theoretical objection! In 1988 the UK Department of Transport for political purposes arbitrarily doubled the value for a human life used in its calculations – this simple stroke of the pen increased the benefit expected from its road schemes by an average of 4.5%, although nothing in fact had changed.

If you remember that the reduced number of deaths will be a guess anyhow, you can see that this sort of exercise is not really worth very much, and the more complicated it gets and the more social benefits that are quantified, the more questionable is the result.

A further difficulty is that by reason of the type of people undertaking these studies the values assigned to non-quantifiables tend to be those of the cultured middle class; the relative values of preserving the environment as against providing local employment, for instance, might not be those which a working family living in the area would choose.

Cost-benefit analysis in its extreme form is now largely discredited, but its successor in the public domain, option appraisal, draws upon its techniques. Option appraisal can be applied to any proposal for public investment, and involves the appraisal of all possible options (including the 'do-nothing' option). Cost-benefit analysis is used to evaluate those aspects which have a clear money value, both of a capital and recurrent nature, but intangibles are merely assessed and shown separately. It is left to the administrators to make the subjective decisions, knowing the financial outcome of the more tangible parts of each option.

The client's needs

All building clients will have a set of needs, some of which are more important to them than others. We have to look at these carefully, because it is common to be told, for instance, that a low cost is required, or that time is important, without these very basic requirements being defined in more detail – and it is the detail that decides the best way of tackling the project. So we will now look at the main time and money requirements which the client may have, and the different forms which each may take, remembering that some clients may have more than one requirement under each heading.

Time requirements

- *No critical time requirements.* This is quite common, especially in social development.

- *Shortest overall time, from the inception of the idea to 'turning the key'.* This is likely to be the requirement on a simple profit development.

- *Shortest contract period from the time the builder is appointed.* This by itself is not often relevant to the client's needs, but it is surprising how often it is asked for.

- *Shortest contract period from the time that construction actually starts on site.* This is a reasonable requirement where, for example, there is a delay in acquiring the property, or where people have to be moved out of property on the site before demolition can take place, or where the building work will cause inconvenience or disruption.

- *Early start on site.* This may be required where the payment of a grant or subsidy depends upon work having been started by a certain date. It is also sometimes asked for by a lazy or incompetent architect to give the client the impression that something is happening at last.

- *Reliable guaranteed completion date stated by contractor.* This may be wanted so that firm arrangements can be made well in advance for commissioning the building.

- *Firm completion date stated by client.* This may apply where the client is under notice to quit existing premises or where there is a particular event that the building must be open for, such as the beginning of the summer season for a hotel.

- *Early completion unwelcome.* It is often wrongly assumed that if a client says that time is critical then early completion will be welcome. This is not necessarily the case; if the client's arrangements are being made on the basis of a particular date, early completion simply means that the client's money has to be wasted watching and maintaining an empty building, and also that the building has to be paid for earlier than anticipated.

- *Phased programme to fit in with plant installation.* This is especially important in the case of sophisticated projects such as TV transmitters or chemical works where the actual building work is only a small part of the total scheme.

- *Handing over in sections.* This is often very important in alteration works or in rebuilding, where people or processes from one section have to be rehoused elsewhere before work can be carried out in that section.

Cost requirements

- *No critical cost requirements.* This is not very common, but it can occur where the building is only a part of a major development project (e.g. the TV transmitter already mentioned) or where the first consideration is quality.

- *Low total cost of whole project.* By contrast, this is almost always said to be the main priority. However, very often one or more of the following criteria are actually the real ones.

- *Low cost of building contract.* The concern here is to keep the lump sum building cost to the minimum, even if this does not minimise financing costs, administrative and supervisory costs, or costs of furnishing and maintenance. It is still quite often required on public-sector projects where these things may come out of different funds.

- *Low cost in relation to units of accommodation.* This is a very usual cost requirement, in both the profit and social sectors.

- *Good budgetary control of the project.* In many cases it is important that the final cost should be as close as possible to the initial cost forecasts, even if this is not necessarily the lowest cost that the competitive market might produce at the time the actual orders are placed.

- *Good forecast of cost at contractual commitment.* This is required by clients who want an accurate forecast of final cost before committing themselves to major expenditure.

- *Best combination of capital and maintenance costs.* One would like to think that this was more common than it is – there are all sorts of reasons like taxation, grants, cost yardsticks, etc., which tend to prevent these two things from being weighted equally (see Chapter 7).

- *Low capital cost.* A more usual requirement, especially if the building is going to be sold, or if running costs come out of a different fund.

- *Low maintenance cost.* Less usual, but may be required if maintenance is going to be inconvenient, for example by putting the building out of commission while it is going on.

- *Timing of cash flow.* This may be required in order to optimise the cost on a discounted cash flow basis (see Chapter 6) or else to phase in with the availability of the client's funds.

- *Minimum capital commitment.* This would be required if the client wanted the contractor to bear most of the cost until the building was handed over.

- *Share in risk of development.* A variation on the last, where the contractor is paid by a share in the profits; this has been used on large speculative developments.

The client, of course, may wish to combine three or four or more requirements from the above list; as a result there are many different sets of possibilities, and the way the project is undertaken should reflect the client's individual priorities and combination of needs. A standard solution should not be adopted simply because it is the usual one.

Summing-up

In the case of profit development:

- The only purpose of the development is to make a profit, and profit is quantifiable.
- If an adequate profit cannot be foreseen the development will not be undertaken, however much it may be needed.
- Once a final decision has been made on income levels the cost will have been determined and must not be exceeded. If any item goes up in cost the money must be saved elsewhere.
- A misjudgement of either the costs or the expected receipts of a scheme will have exactly the same effect on its profitability; neither is more important than the other.
- If the expected profit is not made the project will be a failure from the client's point of view, however pleasant or useful the resulting development may be.

In the case of social development:

- The cost is not a clear-cut measurement of the effectiveness of the project (as it is with profit development); the benefits of a hospital, clinic, school or police station are largely unquantifiable.
- Therefore nobody really knows whether they ought to be spending twice as much money (or half as much) on buildings of this kind, and no amount of cost planning is going to give them the answer.
- The most that cost planning can do is to help use the total allocated funds more effectively within the current framework of rules, and accept that the basic values will be decided for political reasons.

Chapter 5
Money, Time and Investment

Present and future values

If you were asked 'Would you rather be given £1,000 now, or in 5 years time?' you would almost certainly say 'Now', and even if you were living in a time of zero inflation you would still be right to prefer to have the money right away. After all, you might be dead in 5 years time, or there might be a revolution or Third World War before you had the opportunity to collect. Alternatively, the person who was going to pay you might have died, or disappeared, or gone bankrupt, or forgotten their promise. Even if you were not going to need the money for 5 years and were not worried about risks you would still do better to have it now and place it in a savings bank where it could accumulate interest.

We can therefore see that a sum of money at some time in the future will always be worth less than the same amount of money today, and the difference will depend upon:

- the length of time involved;
- future risks;
- the probable interest rate.

In doing the calculations it is a good idea to assume an interest rate that would reflect likely inflation and any special risks over the period concerned rather than a rate which might actually be obtainable today. And, of course, just as a future lump sum is worth less than its equivalent today so are future recurrent expenses or receipts.

If you had to put a sum of money aside to pay somebody £1,000 a year for 10 years the amount required would be much less than £10,000, because the money which had not yet been paid out would be earning interest each year and in the early years in particular this would be quite a lot of money. In fact if interest rates were as high as 10% the £10,000 would provide £1,000 a year for ever, not just for 10 years. The actual sum which would be needed to provide £1,000 a year for 10 years is the amount which, with all its interest earnings, would be exactly used up at the end of 10 years when the last payment is made. Its actual worth would depend upon current interest rates.

Conversely, of course, a sum of money in the past is worth more than the same sum today; £1,000 invested 10 years ago would be now be worth two or three times that amount.

Construction costs in relation to time

Until about 50 years ago it was the usual practice for quantity surveyors to consider the cost of the building irrespective of time, and this was possibly good enough because:

- interest rates were 3% or lower;
- inflation was non-existent;
- buildings were comparatively simple and construction times were quick;
- the techniques of cost estimating were crude anyhow.

Today the situation is quite different:

- Interest rates will continue to be a major factor, especially as rates tend to go up during expansionary periods when large programmes of building work are being undertaken.
- Although inflation is now under control it is far from negligible.
- With the increasing complexity of large buildings it is not uncommon for some years to elapse from the start of expenditure on the project to the time when an income is produced or the building is available for use.
- Clients now expect a sophisticated cost planning service.

Financing charges

'Financing charges' are the interest which the client must pay every month on expenditure to date, until the building is producing an income or is ready for use. An apparently low total cost may not be much of a bargain if it involves a long-drawn-out contract with high early expenditure on which financing charges are piling up. An extreme example of this concerns a very expensive office block in Australia which met with many difficulties and delays in construction during the 1960s and 1970s, and whose financing charges alone came to almost as much as the total building cost of the notoriously expensive Sydney Opera House – although unlike the Opera House costs they did not get into the media!

It may be asked 'Do financing charges apply to all development, or only where profit is involved?' Basically they apply to all development, because the money is either being borrowed and attracting interest charges, or is the developer's own money, which because it has been spent on the development is not available for investing elsewhere and is therefore not accumulating interest. Strangely enough, although most cost planners are aware of this factor, some clients' accounting systems do not recognise it and so in practice cost planners are not always able to demonstrate any benefit from optimising payment in relation to time. However, private developers in particular are certainly aware of the importance of the timing of income and expenditure. And building contractors have always recognised its significance, even if they have not always been able to formalise their methods of dealing with it.

Methodology

In considering development finance we have three kinds of expenditure/income which we need to compare with each other:

- lump sums today;
- lump sums in the future;
- sums of money occurring at regular intervals during the period under consideration (wages, rents, etc).

We cannot compare these, one with the other, unless we modify them in some way in order to put them on a common basis. There are two basic methods, and as usual they are just different ways of expressing the same thing:

Present-day value

All expenditure is expressed as the capital sum required to meet present commitments plus the amount which would have to be set aside today to provide for future payments, discounted to allow for accumulation of interest. Income is similarly treated; future income is discounted to the present day in the same way.

Annual equivalent

This is the total of:

- any regular annual payments and income, such as wages, rents, etc.;
- annual interest on items of capital expenditure;
- a 'sinking fund', the amount which would have to be put away annually to repay the capital cost at the end of the period.

Alternatively, the annual interest and sinking fund can be combined and expressed as the annual instalments which would be required to pay off the capital costs and interest over the term of years in question (rather like paying off house purchase through a mortgage).

Both of these methods will in the end give a similar answer, and which one is used is purely a matter of convenience and depends upon whether you are thinking mainly in terms of capital finance or in terms of annual income and expenditure.

Calculations

It is usual these days to undertake the necessary calculations for these comparisons by using computer programs, or using formulae with the aid of a pocket calculator, rather than by looking up the values in books of tables. The formulae are set out in Appendix B, but are repeated here. Short tables showing a few principal rates of interest are also included in Appendix B to assist those who find them more convenient than doing the calculations.

In the following formulae n represents the number of periods and i the interest rate expressed as a decimal fraction of the principal, e.g. 5% = 0.05.

Calculations are often required that involve monthly, or even weekly, payments and in such circumstances annual interest rates are not of much use. Such calculations should use an equivalent rate for the period, dividing the annual rate by 12 or 52. Strictly speaking this is not correct, as it ignores the effect of weekly or monthly compounding. The exact annual equivalent of a monthly interest rate i is not $12i$ but $(1+i)^{12} - 1$. The yearly equivalent of 1% per month is therefore 12.68%, and conversely the monthly equivalent of 12% per annum is 0.94888%. Where comparisons between two or more alternatives are being considered this difference is rarely of much significance, but it should be allowed for if specific annual interest rates are an important factor.

Formula 1 – Compound interest $(1+i)^n$

If a sum of money is invested for a number of years it will have earned some interest by the end of the first year. Compound interest assumes that this earned money is immediately added to the principal and re-invested on the same terms, this process being repeated annually. Many forms of actual investment provide for doing this automatically, but it is in any case the correct method to use when making investment calculations, since it should not be assumed that money would be allowed to lie idle. Note that this formula is also useful for extrapolating inflation rates over a period of years.

Example

What will be the value of £5,500 invested at 9% compound interest for 5 years?

Formula $(1+i)^n = (1.09)^5$ and shows that £1 so invested will grow to £1.54. £5,500 will therefore grow to 5,500 × £1.54 = £8,470.

Formula 2 – Future value of £1 invested at regular intervals $[(1+i)^n -1]/i$

Instead of a single lump sum being invested we might put away a regular annual amount on the same basis. At the end of each year the total in the fund would comprise all previous investments, plus the compound interest earned on them, and to this would then be added the next annual contribution.

Example

If £150 is invested annually at a rate of 11% compound interest for 6 consecutive years what would the fund be worth at the end of the sixth year?

Formula $[(1.11)^6 -1]/0.11$ shows that £1 annually so invested will grow to £7.91, assuming that the investment is made at the end of each year.

If the investment is regularly made in the earlier part of the year then 11% per annum interest will require to be added to £7.91 for the additional part of the year involved. If the investment were always made at the beginning of each year then a whole year's interest would have to be added, that is, 11% of £7.91 = £0.87. The total per £1 at the end of the sixth year would therefore be

$$£7.91 + £0.87 = £8.78$$

The example concerned a figure of £150 annually and the result for £1 must therefore be multiplied by 150 in each case.

The answer is therefore £1,186.50 if the money is deposited at the end of the year and £1,317 if it is deposited at the beginning.

Formula 3 – Present value of £1 $1(1+i)^n$

In the compound interest example it will be remembered that £5,500 invested for 5 years at 9% compound interest grew to £8,470. The converse of this is that the present value of £8,470 in 5 years time at 9% interest is £5,500, i.e. that is the amount which would grow to that sum at the end of 5 years.

To avoid having to work out present values backwards like this we use the reciprocal of the compound interest formula. One pound in 5 years time at 9% will therefore have a present value of $1/(1.09)^5 = 65.0p$, so £8,470 × 65p = £5,505. The small error of £5 is due to the value being taken to the nearest tenth of a penny only. We shall discover later that the calculations for which these formulae are used are so conjectural that such an 'error' is of no real significance.

Example

What is the present value of £1,200 in 35 years time discounted at 15% per annum?
By use of the formula the present value of £1 in such circumstances is 0.8p. The present value of £1,200 is therefore 1,200 × 0.8p = £9.60.

With £1200 in 35 years time being worth less than £10 today in the example we can understand why there is little point in doing calculations over a period of 40 or 60 years when interest rates are high.

Formula 4 – Present value of £1 payable at regular intervals $[(1+i)^n - 1]/[i(1+i)^n]$

Just as the previous formula showed the present value of future lump sums, this formula shows the present value of future regular periodic payments or receipts over a limited term of years. It is, therefore, very useful for assessing the capital equivalent of things like running costs, wages or rents.

Example

What is the present value of £1,200 payable annually for 10 years assuming an interest rate of 8% per annum?
We see from the formula that the present value of £1 paid annually in such circumstances is £6.71. The present value of £1,200 annually is therefore 1,200 × £6.71 = £8,052. This is the sum which would have to be invested today at 8% compound interest in order to discharge such an obligation and leave nothing over at the end.

It is interesting to see what difference would result if the money were to be paid in quarterly instalments of £300 instead of at the end of each year. Eight per cent per annum is equivalent to 2% per quarter, and (according to the formula) the present value of £1 over $10 \times 4 = 40$ periods at 2% is £27.36. The present value of £300 paid quarterly for 10 years is therefore:

300 x £27.36 = £8,208

compared to £8,052 for the yearly payments of £1,200.

The present value of £1,200 payable for 10 years is sometimes referred to as '10 years' purchase of £1,200'. There is little point in using this old-fashioned land agents' jargon, but you might come across it somewhere.

Formula 5 – Annuity purchased by £1 $[i(1+i)^n]/[(1+i)^n -1]$

This is the reciprocal of the previous formula, and gives the annuity (or regular annual payment) purchased by a lump sum payment of £1. At the end of the given number of years the money will be exhausted. It is therefore useful for calculating the annual equivalent of a given present-day lump sum (whereas Formula 4 calculated the present-day lump-sum equivalent of a given annual amount).

Example

What annual saving in maintenance costs over a period of 10 years would justify an increase in capital costs of £90,000, assuming an interest rate of 8%?

The annual equivalent of £1 by the formula in these circumstances is 14.9p. The annual equivalent of £90,000 is therefore $90,000 \times 14.9p = £13,410$. If the saving in maintenance costs exceeds this amount then the additional capital investment of £90,000 would be justified.

Formula 6 – Sinking fund $1/[(1+i)^n -1]$

It will be remembered that Formula 2 showed the lump sum which would result in the future from investing a fixed sum of money each year (or other periodic interval). However, sometimes you want the same information a different way round; you want to know how much you must invest each year to accumulate a certain sum in a certain number of years (for example to provide for replacing some worn-out piece of equipment). The formula for calculating this is the reciprocal of Formula 2 (Future value of £1).

Example

How much must be invested at the end of each year at 7% per annum to amount to £20,000 at the end of 12 years?

We see from the formula that 5.6p has to be invested each year to produce £1 in such circumstances. For £20,000 the figure is therefore:

20,000 × 5.6p = £1,120

In practice, money is rarely put away into a sinking fund. However, it is a valuable economic concept.

Summing-up

It is important that the principles governing these techniques should be understood, as they have to be used extensively in cost studies and are referred to and developed throughout the rest of the book.

Chapter 6
Developer's Cash Flow

Timing of payments and receipts

Project planning and budgeting needs to take into account not merely the total lump sum involved but the timing of the various payments and receipts. This is the concept of 'cash flow':

- There is a positive cash flow when money is received into an organisation.
- There is a negative cash flow when money is paid out.

Cash flow is a valuable concept because it enables us to look at the financing of a project in a more sophisticated way than merely considering lump sum totals of expenditure and receipts.

Discounted cash flow (DCF)

In the preceding chapter we learned about the time/money relationship, and that monies received or spent in the later stages of a project are worth less than similar sums earlier on. If we therefore discount all the payments and receipts to a common date we shall be able to measure the true profitability of the project. There are basically three ways in which this can be done:

- rate of return;
- value at criterion rate of return;
- payback.

Rate of return

This involves calculating the 'rate of return' (sometimes called 'internal rate of return') on the working capital, and could be used to show the profit as a percentage of average working capital. This has to be done with a computer package, because it involves 'trial-and-error' experiments with different discount rates until a rate is found which discounts the total receipts and the total expenditure to exactly the same present-day figure. It therefore cannot be demonstrated in this book. It is the only DCF method which shows the profit as a percentage on capital employed, but:

- It cannot be used for comparisons where there are no receipts to set against expenditure.
- It can give a misleading result in studies with large short-term positive cash flows, because in practice it would be difficult to obtain substantial returns on these unless such money can be profitably employed in the client's business. Such projects are better dealt with by the 'NPV at criterion rate of return' method, which is next to be described.

Net present value (NPV) at criterion rate of return

This is a much easier calculation, and would probably be more often used by cost planners in assessing the viability of development. It involves the selection of a 'criterion' rate of return, which would, depending on the circumstances of the case, be either the rate at which money can be borrowed to finance the project or a minimum acceptable percentage of profit. The income and expenditure are both discounted at this rate and the difference found. This is the NPV or net present value of the project. If the discounted income exceeds the discounted expenditure then the NPV is positive and represents a profit over and above the criterion rate. If the discounted expenditure is larger than the discounted income then the NPV will be negative. The project would therefore not service its loan or pay the minimum acceptable profit and is unlikely to be carried out.

It will be realised that this method can be used for two quite different purposes:

- to assess whether a project is worth carrying out, or
- to see which of two or more alternative solutions is the best from the point of view of cost.

Payback

Payback is a third DCF technique which is sometimes used in assessing commercial installations, and which compares the time taken to repay the costs of the schemes in question. It is not a very meaningful way of validating building projects, and is rarely if ever used by cost planners. There will be no further reference to it in this book.

Criterion rate of return method used for comparison purposes

When used for comparative purposes there need not be any income to set against expenditure, but unlike the rate of return method the criterion rate method will work in these conditions. If there is no income the calculations will of course always show a negative balance, and the scheme with the smallest negative balance will be the best choice. Instead of using the system to show the NPV it can be used to show the 'annual equivalent' where this is more relevant, as shown in the following example.

A simple worked example

We can now proceed to work out a very simple comparison to demonstrate the system, and to show the use of the formulae given in the previous chapter. As the example involves comparing recurrent expenditure and capital costs both the present-day value and the annual equivalent method are equally appropriate, and it is worked out both ways to show that the two methods obtain the same result.

Example

Comparison of cost of providing a mechanical stoker for a boiler with the cost of employing a fireman, assuming the life of the installation is 20 years and the interest rate is 8% per annum.

1. Present day value
    (a) Fireman's wages, etc., say £12,000 pa
         3 men (three shifts per day) = £36,000 pa
         £1 for twenty years at 8% = £9.82 (Table 4 in Appendix B)

         $\qquad\qquad\qquad\qquad\qquad\qquad$ 36,000 × £9.82 $\qquad$ £353,520

    (b) Cost of mechanical firing equipment,
         together with consequent alterations
         to boiler house $\qquad\qquad\qquad\qquad\qquad\qquad\qquad\qquad$ £300,000
         Total saving by using mechanical firing $\qquad$ £53,520

2. Annual equivalent
    (a) Firemen's wages, etc. (as above) $\qquad\qquad\qquad\qquad$ £36,000
    (b) 8% interest on cost of mechanical
         firing equipment, etc. (£300,000) $\qquad$ £24,000

    (c) Sinking fund to repay capital cost of
         the equipment at the end of 20 years
         (taken at 8% although it is common
         practice to adopt a lower interest rate
         for this purpose) £300,000 at 2.2p $\qquad$ £6,600
         $\qquad\qquad\qquad\qquad\qquad\qquad\qquad\qquad$ £30,600 $\qquad$ £30,600

         Annual saving by use of mechanical
         firing $\qquad\qquad\qquad\qquad\qquad\qquad\qquad\qquad\qquad$ £5,400

Note: Table 5 in Appendix B can be used to combine stages (b) and (c).

3. Check
    £5,400 for 20 years at 8% (Appendix B Table 4) $\qquad$ £53,028
    approximately the same as the capital saving of £53,520

Note that in order to keep the example simple, various extra factors (like maintenance, different type of fuel, etc.) have been ignored.

A further example

A simplified investment project has been chosen as an example.

Example

A plot of land is being bought for £800,000, a block of flats is to be erected, and the whole development disposed of for £2,700,000. In Case A full drawings and bills of quantities (BQs) are to be prepared and tenders invited for a building contract of 15 months' duration at a cost (including professional fees) of £1,320,000. In Case B the project is to be started before full documentation has been prepared and the building programme will be compressed into 10 months. However, the result of this rush will be a cost (including fees) of £1,480,000.

It is required to examine these two programmes to see which would be the more profitable. The cash flows are set out in Table 3. Suppose that a criterion rate of 12% pa is chosen. This will be equivalent to 1% per monthly period and so the two alternatives can be discounted using the 1% DCF tables (Table 3 in Appendix B). It will be seen from Table 4 that both schemes would be profitable assuming 12% per annum as a criterion rate of return, although the profit of Case A would have a present value of £216,080 while the profit of Case B would be slightly lower at £196,980. In these circumstances it would appear to be more profitable to go for the longer and less expensive contract arrangements. If, however, the criterion rate of return was 15%, the NPVs would be reduced to £140,320 (Case A) and £147,540

Table 3   Undiscounted cash flows.

		Case A	Case B	
End of month:	1	–800,000	–800,000	(land purchase)
	3	—	–40,000	
	4	–40,000	—	(professional fees)
	5	—	–40,000	
	6	—	–60,000	
	7	—	–100,000	
	8	–40,000	–140,000	
	9	—	–180,000	
	10	–40,000	–220,000	
	11	–60,000	–180,000	
	12	–60,000	–160,000	
	13	–60,000	–140,000	
	14	–80,000	–120,000	(Payments to
	15	–80,000	–100,000	builder
	16	–100,000	—	and professional
	17	–100,000	—	fees)
	18	–120,000	—	
	19	–140,000	—	
	20	–120,000	—	
	21	–80,000	—	
	22	–80,000	—	
	23	–60,000	—	
	24	–60,000	—	
		–2,120,000	–2,280,000	
	Sale	2,700,000	2,700,000	
		(month 24)	(month 15)	
	Profit	£580,000	£420,000	

Table 4 Cash flows (discounted).

End month:	Case A		Case B	
1	$-800,000 \times 99.0p =$	$-792,080$	$-800,000 \times 99.0p =$	$-792,080$
3			$-40,000 \times 98.0p =$	$-38,820$
4	$-400,000 \times 96.1p =$	$-38,440$		
5			$-40,000 \times 95.1p =$	$-38,060$
6			$-60,000 \times 94.2p =$	$-56,520$
7			$-100,000 \times 93.3p =$	$-93,260$
8	$-40,000 \times 92.3p =$	$-36,940$	$-140,000 \times 92.3p =$	$129,280$
9			$-180,000 \times 91.4p =$	$164,580$
10	$-40,000 \times 90.5p =$	$-36,220$	$-220,000 \times 90.5p =$	$199,160$
11	$-60,000 \times 89.6p =$	$-53,780$	$-180,000 \times 89.6p =$	$-161,340$
12	$-60,000 \times 88.7p =$	$-53,240$	$-160,000 \times 88.7p =$	$-142,000$
13	$-60,000 \times 87.9p =$	$-52,720$	$-140,000 \times 87.9p =$	$-123,020$
14	$-80,000 \times 87.0p =$	$-69,600$	$-120,000 \times 87.0p =$	$-104,400$
15	$-80,000 \times 86.1p =$	$-68,900$	$+2,600,000 \times 86.1p =$	$+2,239,500$
16	$-100,000 \times 85.3p =$	$-85,280$	NPV =	£196,980
17	$-100,000 \times 84.4p =$	$-84,440$		
18	$-120,000 \times 83.6p =$	$-100,320$		
19	$-140,000 \times 82.8p =$	$-115,880$		
20	$-120,000 \times 82.0p =$	$-98,340$		
21	$-80,000 \times 81.1p =$	$-64,920$		
22	$-80,000 \times 80.3p =$	$-64,280$		
23	$-60,000 \times 79.5p =$	$-47,720$		
24	$+2,640,000 \times 78.8p =$	$+2,079,180$		
	NPV =	£216,080		

(Case B). Both schemes are still profitable, but with higher interest rates the shorter and more expensive scheme now shows the better profit. In Fig. 6.1 the NPV and rates of return for the two schemes are shown graphically.

It will be seen that Case A is much the more profitable at low interest rates, that both schemes show an equal NPV of about £164,000 at 14%, and that Case B becomes relatively more and more attractive as the required rates of return rise above this point. Case A ceases to be profitable if a rate of more than 21% is required, and Case B at 25%. This demonstrates the point that if a high rate of return is required it is usually worthwhile to compress the design/building period even if this results in a less efficient building process and higher cost.

Care required with DCF techniques

These discounting techniques are very useful in enabling a project to be looked at from the point of view of profitability. By themselves, however, they only tell part of the story and it is vital that the cash flow situation should be considered as a whole before a decision is made. For this purpose the undiscounted figures for each period should also be looked at, either in tabular form as Table 3 or, better still, in diagrammatic form as in Fig 6.2.

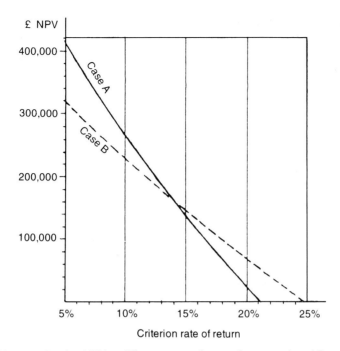

Fig. 6.1   Diagram showing NPV at different rates of return for cases A and B.

In looking at the figure and diagrams we can see that Case B not only requires a higher capital commitment, but also requires it over a much shorter period. By the end of the first year Case A requires an investment of £1,040,000 whereas Case B requires no less than £1,920,000, and the client may not be able to raise this amount of money in the time available. From this point of view Case A might suit a client who was committed to a number of developments in progress simultaneously, while a client who liked to finish one development at a time and then go on to the next might prefer Case B. This question of keeping capital expenditure to a minimum, or of phasing payments, might be more important than a small difference in NPV.

A further factor which could offset any advantage which Case B might have over Case A would be the inflation of property prices. Since the price of houses and flats in many parts of the country has been increasing more rapidly than the general rate of inflation it is possible that the flats which would sell for £2,700,000 at the end of 15 months might sell for £3,000,000 at the end of 24 months, so destroying the previous calculations and giving an advantage to Case A.

It is worthwhile remembering that undiscounted cash flow statements and diagrams can be very useful in the social development field, where the funding of capital expenditure may be done on a quarterly or yearly basis, and it is useful to be able to plan which period the expenditure will fall into. This topic is developed in greater detail in Chapter 20.

While DCF appraisals are useful, they must be used with discretion. In any case, DCF assessments of investment can only really work if there are quantifiable receipts to set against expenditure; otherwise they merely prove that the less

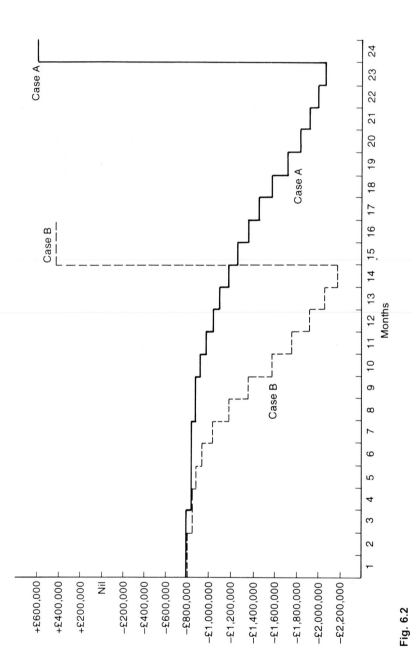

Fig. 6.2

that is done, and the slower it is done, and the longer it can be put off, the better. Unlike straightforward cash flow statements, DCF therefore has little place in programming a purely social development, although it can be used for cost comparisons of alternative solutions within a given time scale.

Summing-up

Discounted cash flow (DCF) techniques are very useful for investigating the profitability of projects, and for comparing alternatives. The criterion rate of return version is particularly suitable for use by cost planners, as it can be used manually and can also be used for comparing alternatives on non-profit projects. But DCF studies do not tell the whole story, and it is important to look at undiscounted figures to see if they reveal any snags.

Chapter 7
Whole-Life Costs

Basis for whole-life costing

Cost targets for projects have traditionally been expressed only in terms of initial capital cost, and many people think it is about time that this changed, because once the building has been handed over and paid for it will continue to cost money – it will need to be maintained, decorated, heated, cleaned, repaired, and so on – and from time to time during its life quite expensive refitting may be required. And in order to meet tight initial cost targets, materials or constructions may have been used which are low in first cost but which require constant expenditure on maintenance and repair during the life of the building. If only, it is argued, the architect were allowed to spend more money initially on better materials these would pay for themselves in the long run.

The term 'whole-life costing' is used to describe a form of modelling technique to cope with this mixture of capital and running costs. 'Life-cycle costs' is an older term for the same thing, 'costs-in-use' is an obsolete one.

A simple illustration

As a simple example let us consider the walls of a lavatory in a good class building. To plaster the walls and paint three coats of gloss paint would cost about £12.00 per square metre, whereas glazed tiling would cost £40.00. As the tiling would last 50 years with little attention, and the plaster walls would need to be painted every 5 years or so at a cost of £4.00, the total cost at the end of 50 years would be:

- £48.00 for the plaster walls (£12.00 + (9 × £4.00)) and
- £40.00 for the tiling.

Apparently, therefore, the better and more hygienic finish would cost less than the inferior work.

Similarly with roofs; a sound lead roof will last up to a hundred years and will then have quite a considerable scrap value whereas a built-up felt roof might have to be renewed three or four times during this period. The difference of 400% in the initial cost of these two types of roofing would appear to cancel out by the end of the century, and the trouble and inconvenience of having the roof stripped and relaid at intervals would be avoided by using the more durable material.

Of course, such examples are oversimplified, because we know that in comparing current and future expenditure we have to consider not the actual amounts to be spent in future years but these amounts discounted to present day value – that is, the sum of money which would have to be invested today in order to accumulate to these amounts by the time they are needed. As money invested at as little as 5% compound interest doubles in just under 15 years there is obviously a considerable difference between the two figures.

Components of whole-life costs

Whole-life costs have three components:

- capital sums, now or in the future;
- recurring costs (running costs, maintenance, etc.);
- sinking fund (to repay the capital when the asset is life expired).

A worked example

At this point we can look at a simple example (the present day values are calculated according to the formulae in Appendix B). There is no need to bother with a sinking fund in the example, as this is a comparison only and both claddings will be equally worthless at the end of the 40 years.

Example

It is desired to compare the whole-life costs of two types of cladding to a factory building, whose life is intended to be 40 years. The rate of interest allowed is 3%, per annum compound.

Whole-life costs of Cladding A
Cladding A will cost £1,000,000, will require redecorating every 4 years at a cost of £120,000, and will require renewing after 20 years at a cost of £1,400,000.

	£
Initial cost	1,000,000

Present value at 3% of:

Redecoration after 4 years £120,000	at 88.8p =	106,560
Redecoration after 8 years £120,000	at 78.9p =	94,680
Redecoration after 12 years £120,000	at 70.1p =	84,120
Redecoration after 16 years £120,000	at 62.3p =	74,760
Renewal after 20 years £1,400,000	at 55.4p =	775,600
Redecoration after 24 years £120,000	at 49.2p =	59,040
Redecoration after 28 years £120,000	at 43.7p =	52,440
Redecoration after 32 years £120,000	at 38.8p =	46,560
Redecoration after 36 years £120,000	at 34.5p =	41,400
			2,335,160

Whole-life costs of Cladding B
The alternative Cladding B will cost £1,800,000, and will last the life of the building without any maintenance, although a sum of £300,000 is to be allowed for general repairs after 20 years.

	£
Initial cost	1,800,000
Present value at 3% of:	
Repairs after 20 years £300,000 at 55.4p =	166,200
	1,966,200

Saving by using Cladding B is therefore £2,335,160 minus £1,966,200 = £368,960.

It would therefore appear to be justifiable to use the initially more expensive Cladding B, as this will prove much the cheaper in the long run. Note that the 'present value' method of discounting has been used, as the 'annual equivalent' method (see Chapter 5) would have proved too complicated in such an instance, where none of the payments or receipts are annual.

A quick method

It is worth noting that the present value of Cladding A could have been evaluated even more quickly, with a slight loss of accuracy (because of averaging out the decoration costs to a yearly figure), as follows:

Example

Whole-life costs of Cladding A (quick method)

	£
Initial cost	1,000,000
Redecoration £120,000 every 4 years = £30,000 p.a.	
36 years purchase of £30,000 at 3%	
(see Appendix B Table 4) at £21.83 =	654,900
Replacement at 20 years	
£1,400,000 less saving in decoration £120,000 = £1,280,000	
Present value (3%) at 55.4p =	709,120
	2,364,020

The slight loss of accuracy by this method is of no practical significance, because the assumptions which have been made are just that – assumptions.

Effect of different assumptions

In order to demonstrate the effect of quite small misjudgements of the future, the example relating to Claddings A and B is now worked out again, assuming some slight differences to the assumptions previously made. All these differences lie well within the range of error to be expected with careful estimates in normal times, and exclude either inflation or anything going badly wrong.

Revised example

The rate of interest allowed is 4%, per annum compound, an increase of 1%.

Revised whole-life costs of Cladding A
Assume that Cladding A is redecorated every 5 years instead of every 4 years, at a cost of £100,000 and lasts for 25 years instead of 20, costing £1,200,000 to renew.

	£
Initial cost	1,000,000

Present value at 4% of:

	£
Redecoration after 5 years £100,000 at 82.2p =	82,200
Redecoration after 10 years £100,000 at 67.6p =	67,600
Redecoration after 15 years £100,000 at 55.5p =	55,500
Redecoration after 20 years £100,000 at 45.6p =	45,600
Renewal after 25 years £1,200,000 at 37.5p =	450,000
Redecoration after 30 years £100,000 at 30.8p =	30,800
Redecoration after 35 years £100,000 at 25.3p =	25,300
	1,757,000

Revised whole-life costs of Cladding B
Assume that Cladding B has to be repaired after 15 years and again at 30 years, instead of only once at 20 years.

	£
Initial cost	1,800,000

Present value at 4% of:

	£
Repairs after 15 years £300,000 at 55.5p =	166,500
Repairs after 30 years £150,000 at 30.8p =	61,600
	2,028,000

Saving by using Cladding A is therefore £2,028,100 minus £1,757,000 = £271,100

This calculation gives a completely different result to the original – the initially cheaper cladding proves to be much cheaper in the long run also. It would still have life left in it at the end of 40 years if it were decided to keep the building in commission for a bit longer.

Conclusions drawn

The lesson these examples teach is that calculations of this kind should not be used as the sole basis for decision unless the cost advantage of one of the alternatives is really massive – not the 15–20% of these examples – and is unlikely to be upset if the future turns out to be a bit different to what was assumed. The examples also show how easy it is to manipulate the assumptions in order to give a politically desirable answer.

A mathematical technique known as 'Monte Carlo simulation' exists for assessing the effect of probabilities and might assist decision-making in some circumstances. Alternatively, the same result can be obtained by manipulating the above calculations on the computer, using a number of different values for the variables. This may enable the important ones to be identified.

Inflation

The last examples did not take inflation into account. In the late 1990s inflation in the UK was only 3 to 4% but it was in double figures 10 years previously, and it could just as easily rise again in the future with a change in economic circumstances. Some authorities suggest that this factor really makes complete nonsense of any attempt at long-term forecasting over periods of 20 or 30 years or more. This is to take too gloomy a view; inflation may be no worse than the other factors which we have discussed, because:

- interest rates have to rise during a period of rapid inflation as otherwise nobody would invest when the value of their capital was being eroded;
- rents and other income from buildings tend to rise at roughly the same rate as inflation in the long term (otherwise building owners would go bankrupt).

If, therefore, the original Cladding A example were worked out on the basis of 10% annual inflation and 13% interest we would get much the same result as with no inflation and interest at 3%. (Note that the compound interest table has been used for calculating the effect of inflation.)

Example

Revised whole-life costs of Cladding A (10% annual inflation)

	£
Initial cost	1,000,000
Present value at 13% of:	
Redecoration after 4 years £175,200 at 61.3p =	107,398
Redecoration after 8 years £256,800 at 37.6p =	96,556
Redecoration after 12 years £376,800 at 23.1p =	87,040
Redecoration after 16 years £550,800 at 14.1p =	77,662
Renewal after 20 years £9,422,000 at 8.7p =	819,714
Redecoration after 24 years £1,182,000 at 5.3p =	62,646
Redecoration after 28 years £1,730,400 at 3.3p =	57,103
Redecoration after 32 years £2,533,200 at 2.0p =	50,664
Redecoration after 36 years £3,709,200 at 1.2p =	44,510
	2,403,293

This is very similar to the figure which we had originally, in spite of the vast increase in cost of decoration from £120,000 a time to almost £4,000,000.

Relative inflation costs

The published rates of general inflation are not always a good guide because:

- Building costs do not necessarily increase at the same rate as general inflation.
- Repair and maintenance costs, which have a high site labour content and sometimes involve obsolescent materials, increase more than a general building cost index would indicate.

- Differential inflation may occur where the price of one type of commodity or factor increases at a different rate to the alternative. This could be because of the relative scarcity of different raw materials and other resources in the future.

Disadvantages of whole-life costs assessment

The advantages of the whole-life method of comparing costs are self-evident; it enables us to consider the long-term implications of a decision, and to provide a way of showing the cost consequences of short-sighted economies. Unfortunately there are a number of fundamental disadvantages, which explain why this technique for comparing the cost of alternative materials and constructions has been seen more often in textbooks and in the university examination room than in real life. These disadvantages may be expressed through two lines of argument: (1) that initial and running costs cannot be equated; (2) that the future cannot be forecast.

Argument 1: initial and running costs cannot really be equated

This is because on profit developments:

- Where the building is to be sold the maintenance charges will fall upon the purchaser, and so are of little importance to the developer, who is responsible for the construction costs alone.
- Where the building is to be let, or used commercially, the repair and maintenance costs are deducted from the receipts in calculating profit for the year, and are therefore paid out of income before taxation. However, money spent at construction stage has to be raised as part of the capital cost and is eventually repayable. Repayment of, or interest on, capital expenditure is not normally deductible from receipts for the purposes of tax calculations.

In addition, on social developments:

- Even with publicly owned buildings it is an advantage to pay maintenance out of running costs instead of incurring a heavier capital debt due to high construction costs.

- In some instances, such as schools, the bulk of the construction costs may be paid by one authority while another authority will be responsible for running costs, so that there will be little incentive to provide an unduly high standard of building with a view to subsequent saving.

On all developments:

- Money for capital development is normally more difficult to find, and is subject to more constraints, than money for current expenditure.

- Although a building may still be perfectly sound half-way through its planned life, it may be too old-fashioned in design and accommodation to do the job that is required of it in modern conditions.

- The same applies to expensive but durable finishes and fittings, which although still in good condition may give a very old-fashioned appearance to a building. Old joinery, shopfronts, tiling and other finishes or fittings may be ripped out long before they are life-expired, especially in the competitive world of commerce.

- In comparing figures of increased capital expenditure against future costs of repair and renewal it must be remembered that once the money has actually been spent it is not possible to amend the decision in the light of future developments.

- As a domestic example of this, if very expensive finishes are chosen for a house to save repainting, and after a few years the owner becomes short of money, the annual interest on the expensive house still has to be met, whereas if the owner had opted for a cheaper house the redecorating could have been deferred for a year or two. (The house could also be redecorated in the latest fashion each time, if the owner so wished.) It is questionable practice to restrict the actions of future generations by committing them to high interest and repayment costs; this is just as bad as the other extreme of committing them to inflated running costs and maintenance costs by unduly low standards of design and construction.

- Whole-life cost techniques are unrealistic, because it is most unusual, in practice, to set money aside to meet future expenditure. If, for instance, you had to meet repair bills this year for £90,000 it would be little consolation to learn that somebody had discounted this to £20,000 in the year 1980 – you would still have to find the £90,000.

Argument 2: the future cannot really be forecast

- While present-day capital costs can be estimated quite accurately, the cost of maintenance is a pure guess.

- The amount of money spent on decoration and upkeep of a building is determined far more by the current policy of the body responsible for the maintenance than by any quality inherent in the materials. Some owners will redecorate every few years, mend or replace worn or damaged work immediately and continuously carry out a policy of minor improvements; others will spend the very minimum necessary to keep the building in operation.

- Major expenditure on repairs is usually caused by unforeseen failure of detailing, faulty material or bad workmanship, rather than by predicted overall ageing, and so is almost impossible to forecast. A well designed and maintained piece of cheap construction might last much longer than its theoretical life, while some quite expensive work could require early renewal because of, say, entry of water at a badly designed joint.

- Interest rates, which in general tend to reflect the current minimum lending rate (MLR), cannot be forecast with any certainty, particularly over long periods of 20 years or more; remember that net interest rates are also affected by changes in taxation. Between 1988 and 1990 the MLR, although forecast by the Treasury to remain static or fall, actually increased from 8% to 15%. And that was during a period of only 2 to 3 years – would you like to guess what the Bank of England (or the European Bank) will do in the year 2025?

- During the last 40 years there has been almost continuous economic inflation, and any building or maintenance work is likely to cost several times what would have been estimated, say, 20 years ago.

Practical use of whole-life costing techniques

Although whole-life costing techniques have been available for well over 30 years their use in practice has, as suggested above, been very limited. Some of the reasons for this have already been given, but the most powerful causes are that

- the present capital costs are very real;
- the future maintenance and renewal expenditure is usually just 'funny money'.

It is not generally the practice to set funds aside for future maintenance and renewal at the time the capital decisions are made, and when these costs actually have to be faced it is no good turning back to ancient projections of expenditure for a solution. In addition, remembering that it is easy to make assumptions that will give the preferred answer, and that none of the long-term forecasts are going to come home to roost, it is little wonder that whole-life costing is often viewed with suspicion.

Even where designers genuinely want to base their decisions on life-time costs they are often beaten by the pressures of meeting deadlines and cost targets, and conjectural calculations are relegated to second place.

On the other hand, the works departments of some public and recently privatised organisations are now faced for the first time with the need to consider interest on capital as a factor, and their governing boards are tending to ask for whole-life cost appraisals of projects to help them deal with this unfamiliar situation.

In general, however, whole-life costing is at its least effective when dealing with long-term static structures such as buildings. In fact it can give dangerously misleading answers in really long-term situations. Suppose that two alternative materials are being considered for a new main sewer:

- a traditional material with a life exceeding 100 years, and
- a cheaper material with an estimated life of 60 years.

Even a modest discount rate will discount both renewal costs to almost zero over such long periods, with almost no difference in present value between the 60-year and 100-year materials. The study therefore will be dominated by present-day capital costs, and the cheaper material with the shorter life will always be shown to be the better buy, even on a whole-life basis. However, if you were now responsible for maintaining a sewer that had been laid in the 1940s, the difference between a life of 60 years and a life of 100 years would be very real!

Important uses for whole-life costing

However, there are two fields where whole-life costing techniques work very well, and are increasingly being used:

- In dealing with shorter-life assets, such as mechanical or electrical equipment, where foreseeable energy consumption and maintenance and renewal programmes generate much of the future costs.
- Where both the present and future costs are equally real, for instance in a rolling maintenance programme for a major installation where the money is coming from the same fund and policy can be planned accordingly. If these conditions can be fulfilled then whole-life costing is a must!

Reasons for accepting higher capital costs

The rather obvious link between low first cost and high maintenance costs was mentioned earlier in this chapter, and could easily be taken for granted. However, in spite of quite considerable efforts no evidence has been found from general user experience to suggest that increased capital expenditure on better materials and finishes necessarily produces lower maintenance costs. However, this may only be because the sort of organisations that go in for high-grade materials also have high-grade maintenance standards.

Nonetheless, apart from whole-life cost criteria there are still good reasons why a more expensive alternative may be preferred at design stage:

- For prestige reasons. The clients may consider that their building is too important to have cheap and inferior materials and workmanship.
- As well as being more durable, expensive materials are often more pleasant to look at or to use.
- Replacement or repair may be inconvenient. The recovering of a roof, or relaying of a floor, will upset the users of the building and could have severe commercial effects (customers prefer not to go into a shop or hotel where building works are in progress).
- Replacement or repair may be difficult, and therefore expensive. Repairs to the covering of an external cornice might involve scaffolding the whole building, for instance.
- The saving of money on a specific item may involve repair costs out of all proportion to the saving. The use of iron instead of copper for slate nails would eventually necessitate the stripping and reslating of the whole roof, for the sake of saving a few pence per square metre. Similarly, poor quality

galvanising on steel windows might involve the replacement of all the windows after some years.

- Obsolescence may not be a factor of any importance. Such buildings as churches, or especially cathedrals, employ the best materials in the expectation that they will endure for hundreds of years.

In designing a new building therefore it is important to consider what its economic life is likely to be, and what changes it is likely to undergo during that life, before coming to a decision on construction and finishes. Obviously this can only be done to a limited extent, as it is impossible to look too far into the future, but some types of building are inherently more subject to change than others, and this can be taken into account. Theoretically it might be possible to design a building so that all its components would begin to fail simultaneously at the end of its economic life; in practice this is not possible although the worst incongruities can be avoided. Expensive materials should not be chosen for the sake of their durability if they are likely to outlast the rest of the building, or if they will have to be replaced due to the failure of some other and inferior component.

Whole-life costs of mechanical and electrical (M&E) installations

As previously mentioned, this is a rewarding field for the application of whole-life costing techniques, because:

- The proportion of building cost represented by M&E work is continually increasing.
- The conflict between capital outlay and running costs occurs in a pronounced form in the evaluation of energy-consuming systems. Considerable economies in daily fuel and staffing costs can be achieved by additional capital expenditure.

The disadvantages of whole-life costing calculations which were set out earlier in this chapter will not apply so severely as in the case of the building fabric. This is because:

- The running costs of energy-consuming systems form a high proportion of their total whole-life costs (for this reason it is usually convenient to use the annual equivalent method for the whole-life cost study rather than the present-day value).
- The fuel consumption and staffing requirements (e.g. boiler-man, lift attendant) of different kinds of systems can be calculated quite accurately, and cannot be so easily varied by client policy as can general building maintenance and decoration.
- Mechanical and electrical installations should be written off over a much shorter period of time than a general building fabric, both because of the more limited life of their components and because they become obsolete

much sooner. Not many 50-year-old buildings still have their original heating and electrical systems.

- Since forecasts are being made over a shorter period, any assumptions about cost, interest rates and taxation are more likely to be valid.

The most simple forms of calculation are those related to capital/fuel costs of different heating systems.

Example

Automatic gas-fired heating system

	£
Annual fuel costs	10,000
5% interest on capital cost of system (including associated building work) £60,000	3,000
'Sinking fund' to repay capital cost after 20 years at 3% £60,000 at 3.7p	2,220
	15,220

Electric storage heating system

	£
Annual fuel costs	13,600
5% interest on capital cost £30,000	1,500
'Sinking fund' to repay capital cost after 5 years at 3% £15,000 at 5.4p	1,620
	16,720

Annual saving on use of gas-fired system £16,720 minus £15,220 = £1,500

Note that the annual equivalent method has been used here, because it is the saving in annual costs that is wanted. A more complex calculation would have to be used to take into account the cost of additional thermal insulation of the building, for example double glazing. Such a calculation might have two different objects:

- To justify the capital expenditure on the insulation itself, because of savings in fuel cost.
- To justify as well the use of a low-capital/high-fuel-cost heating system, which would become more economic if the required heat loading could be reduced.

Very complex calculations are involved where the scope of the mechanical services are dependent upon the configuration of the building – for example:

- Heat extraction systems might be necessary for a multi-storey building which has a very low wall/floor area ratio and would have to be taken into account when structural cost comparisons are carried out.
- Relative lift costs would be a factor in trying out alternative arrangements for a tall building.

Finally, in all these comparisons, it is likely that in the end the client's own (possibly subjective) views on acceptable levels of capital cost and running cost would be more important than the exact result of comparing them on an economically equal basis. It is important, however, that the client should be given the facts (including these comparisons) before arriving at a decision.

Whole-life costs of the total project

In arriving at a decision on what kind of building to erect a forecast of the whole-life cost of the total project may be important for three reasons:

- To fix rentals of a development, such as an office block or service flats, where the owner will remain responsible for some of the running costs.
- To see whether a client can afford to run the building when it has been erected – as in the case of deciding the optimum size of a new village hall.
- To see whether increased expenditure on a building which is to be occupied by the client, such as a hospital, could reduce the general level of running costs by making it possible to operate with fewer, or less skilled, staff.

In all such cases the problem is usually a short-term one – involving, say, 5-year leases in the first instance and current staffing difficulties in the last. So what is required is a fairly accurate assessment of expenditure in a typical year in the very near future, rather than a 'plus' or 'minus' comparison with capital building cost, because:

- In the first of these cases there is little point in equating annual costs with increases or decreases in the capital cost of the development, as long as the rents will bear the likely charges.
- In the second case, since capital and running costs are almost certain to be met from different sources (perhaps an appeal plus grant for the first, and income from lettings for the running costs), much the same applies.
- It is only in the third case that whole-life comparisons of building costs are likely to mean anything, although even here the main motivation is likely to be staffing problems rather than costs as such.

Information on maintenance costs

Without the existence of a large quantity of published data it would have been difficult for cost planning to take root, and one of the factors making whole-life cost studies so risky has been the lack of any similar body of information on maintenance costs.

In order to remedy this situation a Building Maintenance Cost Information Service (BMCIS) was set up by the RICS, to operate on a fairly similar basis to the earlier established Building Cost Information Service (BCIS). The new service was able to get off to a flying start as all the information which the government had accumulated on its own building maintenance costs was made available to the information service. Full details of this service are given in Chapter 13.

For continued success, however, it is essential for a large number of major property owners to contribute to the store of information, and two difficulties present themselves:

- Rightly or wrongly, such information is often considered commercially sensitive.
- Cost records are usually kept to meet internal costing and budgetary needs, and not in sufficient detail to enable work to individual components and sub-systems to be identified.

Cost-in-use tables were prepared by the now-defunct Property Services Agency (PSA) and were published by HMSO, the most recent version being the third edition (1991). These were particularly directed towards the budgeting of programmes and gave not only costs but risk assessments, under the headings of:

- capital
- cleaning
- cyclical
- repairs

At the time of writing it is not known whether these will continue to be updated by the private-sector organisation which took them over from the PSA when it was disbanded.

Simulated and estimated costs

There is, however, a school of thought which holds that simulated or estimated figures for expenditure are at least as useful as recorded data. This rather surprising view relies on four arguments:

- Data relate to the past whereas simulation relates to the future – which is when the expenditure is actually going to be incurred. Maintenance and servicing costs of a fleet of cars, for instance, recorded over the last 10 years would not be much guide to the next 10 years, because of increases in reliability and corrosion protection, and the increasing use of electronic monitoring and controls.
- You will be very fortunate if historic cost data have been kept in exactly the form that you want them – if you need information on specific items of boiler-house equipment you may well find that costs have only been recorded at the level of 'boiler-house generally'.
- The basis on which cost data are gathered is always suspect – for example figures filled in on time sheets at random on a Friday afternoon, or the use of

wrong job numbers or the possibility that similarly named cost centres from different sources will have differently defined parameters.

- The assumptions made in simulation are known.

Taxation and grants

One major assumption which has to be made in whole-life cost calculations is the degree to which taxation will affect the final results. Taxation is a fluctuating variable, like other assumptions such as building life, component life, inflation and interest rates. Different governments will have different policies towards the amount and type of taxation to be levied and this will directly or indirectly affect the amount the client pays for maintenance.

It is currently possible for a company to write off the cost of most operational and maintenance costs against profit before arriving at the net figure on which corporation tax is payable. In effect this means that the government is giving a very substantial discount on the cost of these items (25–35% in 1998–2000).

In practice, the benefit is probably not quite as great as the tax percentage, because the relief from tax does not usually occur until the next financial year after the costs have been incurred. The effect of this 'tax lag' (which can be as long as 20 months) is to reduce the saving by the discounted value for the time elapsed between payment for the work, and the time at which tax payment is actually due. Hence with a discount rate of 10% and a tax lag of one year, the real benefit becomes:

$$£0.35 \times 0.909 = £0.318 \text{ in every pound}$$

This position becomes more complex when tax relief is allowed on initial costs. For example, to encourage investment in industrial buildings the government has allowed 4% of the construction cost to be written off against tax in the first year, and 4% per year thereafter until the whole of the capital cost has been accounted for. Each of these savings would need to be discounted using the present value table, bearing in mind once again that there is usually some form of tax lag. In addition to tax savings on initial cost it is also possible from time to time to receive direct grants from the government for building in specially designated development areas.

These grants reduce the effective initial capital cost to the client. All these factors should be taken into account when undertaking a whole-life cost exercise. However, it should not be forgotten that the taxation relief is only available on profit and not every firm makes a profit every year.

One last point to bear in mind is the influence of value added tax, now payable on all construction work except qualifying residential accommodation.

Sensitivity analysis

We have already seen that there are a large number of assumptions made in any whole-life cost calculation, and that it is not always possible to access the effect of changes in these assumptions realistically.

One method of testing whether the results achieved by whole-life cost studies are satisfactory for decision-making purposes is to repeat the calculations in a methodical way, changing the value of a single variable (i.e. assumption) each time. It is then possible to see how sensitive the results are to changes in the variable under consideration. If the results are plotted on a graph then a visual check can be made to see when, for example, one component becomes more attractive than another.

Summing-up

The forecasting of running costs of a building will often be useful, but this does not always mean that a whole-life cost study (equating capital costs and running costs) will have much meaning for the client. Such studies are most likely to be justifiable where one, or preferably both, of the following conditions are fulfilled.

- Both sides of the calculation represent 'real' money, and the figures for maintenance and renewal will actually be used for programming, controlling and monitoring the costs of such work.
- The study relates to energy-consuming equipment with a well defined servicing programme.

Even so, the following points should be remembered:

- The shorter the period involved the more accurate the forecast is likely to be. As an alternative the use of an artificially high rate for discounting will have much the same effect, by attaching little value to anything which may happen in the very distant future.
- While the results of the study will be of use as one of the factors to bear in mind in arriving at a decision, it would be most unwise to come to a decision regarding alternatives on a whole-life cost basis alone, unless the calculated advantage of one of them is very substantial and a good deal of trust can be placed in the assumptions made.

In undertaking whole-life cost studies it is important to avoid double counting inflation through using a discount rate based on current gross interest rates (which allow for inflation), and then adding a separate inflation allowance as well.

The objectives of whole-life costing are altogether admirable, and it is a pity that there are so many reasons, which have been set out in this chapter, why it often fails to deliver what it promises. However it is currently a very politically correct approach, and it may not always be a good idea to share all our doubts with a client body – or even with a university examiner. It is at its least effective when dealing with long-term static structures such as buildings, and at its best when dealing on a relatively short-term basis with:

- energy-consuming systems and
- building maintenance programmes.

If used intelligently it can often be very useful. You must make up your own mind about this!

Further reading

Ferry, D. & Flanagan, R. (1991) *Life Cycle Costing – A Radical Approach*. CIRIA Report 122, London.
Flanagan, R., Norman, G., Meadows, J. & Robinson, G. (1989) *Life Cycle Costing: Theory and Practice*. Blackwell Science, Oxford.

Chapter 8
The Client's Budget

Total cost

When any type of building development takes place the total expenditure on the development will be much higher than the net cost of the building fabric, and will comprise:

- cost of land;
- legal, etc., costs of acquiring and preparing the site, and obtaining all necessary approvals;
- demolition or other physical preparation of the site such as decontamination, archaeology;
- building cost;
- professional fees in connection with the above;
- furnishings, fittings, machinery, etc.;
- costs in connection with disposal (sale, letting, etc.) where the building is to be disposed of at completion;
- value added tax on above items where chargeable;
- cost of financing the project (this principally represents interest, etc., on the money which has to be spent before any return is obtained either by way of income or of use);
- cost of management, running and maintenance where the building is to be retained by the client for use or only partially disposed of, or where the building is let to a tenant but the owner has accepted responsibility for some or all of these costs.

In commercial development there is a very close relationship between how much the project will produce in revenue and its commercial value, the latter being largely the capitalised value of the former, and the total costs of the scheme must not exceed this commercial value. In social development this relationship may be expressed by quantifying what are considered to be social benefits, but the issue is never so clear-cut and other criteria may be more important in assessing the amount that can be spent on the scheme.

Building cost

It is important to realise that 'building cost' is only one of the ten items of the client's total cost, and it may be thought that many cost planning exercises are unduly obsessed with this one item rather than with the whole package. There is some truth in this allegation, which is probably due to the fact that many of the techniques were developed in the public service in the days when building costs were closely examined but total costs were rarely considered.

Today this is rarely the case. In extreme circumstances, indeed, keeping the 'building cost' down may not be particularly important. For example:

- The cost of the land may be such an important factor in the costing exercise that everything else has to be fitted round it and the problem then is to get the maximum amount of accommodation that the planning authorities will allow, almost regardless of cost efficiency.
- Alternatively, the building may be only a part of the total project, as in the setting up of a TV transmitter station where the cost of the mast, electronic equipment and cable laying dominate the project and the transmitter building is of minor importance. In these circumstances it would be inappropriate to consider the cost of the building in isolation – it is much more important that it should phase in with the general programme and that its completion should not be a source of worry to the engineer in charge. It may not be necessary to indulge in cost planning of the building at all.

In the later stages of the cost planning process building costs will dominate the calculations. At early budget stage, however, they can only be estimated very generally, because so little is known about the project.

Land values and development

One of the most important items on any developer's balance sheet is the cost of the land and we must consider this aspect first.

Building development of any kind requires a plot of land on which it can take place, and which, once used, will no longer be available for any other development, unless the first one is either demolished or converted. This makes building development quite different from other enterprises, because no developer can start work until the firm or organisation has acquired an interest in the land on which the development is to stand. This has all sorts of economic consequences, because the supply of land is largely fixed. As a now famous estate agent's advertisement put it:

'Buy now! They have stopped making it!'

Like the prices of other commodities, land values are affected by the laws of supply and demand. However, unlike other commodities the amount of land available in any one area is finite and cannot be increased at times of high demand by manufacturing at a higher rate, or importing from outside. This means that if there is a high demand in a particular area the prices of land in the locality will rise very steeply, even though similar land may still be cheap 50 miles away.

Role of the valuer

Valuation surveyors are the experts on the value of land or buildings for investment and the income which various types of development can be expected to produce, but anybody who is involved in cost planning of buildings needs to have some knowledge of the factors affecting the costs of land for development.

Development value

The 'development value' of a piece of land is the difference between the cost of erecting (or converting) buildings on it and the market price of the finished development including the land. Nobody can commission building operations on a piece of land unless they have a financial interest in it; if possible they would wish to own it, so as to obtain the full benefit of the 'development value', although in some areas where land is in very short supply they may have to lease it.

The development value of the land will be determined by:

- Its location – both its geographical region and its local position in that region in regard to amenities, communications, etc. Real estate agents often emphasise the importance of this by saying that 'the three most important factors in land value are location, location and location'.
- Restrictions on its use, imposed either by the vendors in the form of covenants or by the community in the form of planning restrictions.
- Any 'easements' which go with the land, such as right of way across it.
- The physical state of the land – whether it is level or very hilly, and whether there are buildings on it which need to be demolished.
- The current state of the economy.

To illustrate how the value may be determined by position, we can do a simple analysis showing the cost effect of a family house in an inner London suburb compared to a similar house in the outer commuter area, from the point of view of somebody employed in London.

Example

Annual costs: London house	£
Building society repayments	11,500
Daily travel expenses to City	500
Travelling time say $1\frac{1}{2}$ hours per day	
200 days per annum = 300 hours at £10	3,000
	15,000

Annual costs: similar outer area house	
Building society repayments	4,000
Annual season ticket to City	2,500
Travelling time say 4 hours per day	
200 days per annum = 800 hours at £10	8,000
	14,500

Annual saving of outer area house is £15,000 minus £14,500 = £500

The annual costs of the two alternatives are fairly equal, and it is this calculation which helps to explain the different prices of two quite similar houses whose actual building costs would be more or less identical.

Effect of building cost on land value

We can express the value of land for development as an equation:

Value of development = price of undeveloped land + building costs + profit

As we have seen, the value of the development in a free market is its worth to consumers (as compared to other things they can spend their money on), and this in turn largely fixes the price of the undeveloped land. The equation can therefore be better expressed as:

Price of undeveloped land = value of development – (building costs + profit)

The people selling the undeveloped land should be just as capable of doing this calculation as the developer is, and will fix their selling price accordingly. It is therefore not really correct to say that the cost of land pushes up the price of housing or other accommodation; in fact it is the other way around – the market price of development pulls up the cost of land.

In conditions where there is a shortage of suitable building land and a constant rise in development values, as in the south-east of England, fluctuations in building costs will have little effect on land prices. However, if the situation were reversed and development values remained constant or even fell, then an increase in building costs would have the effect of reducing undeveloped land values. It is normally the price of 'undeveloped land' which is the result of the equation, when the other figures have been filled in.

Social considerations of land use

Most people agree that a completely free market in land for development is socially undesirable, and the setting of limits on the scale and nature of development under town and country planning legislation brings a measure of public control to the process. Even so, there is a strong body of opinion that wants to see the public obtain much of the benefit of any increased development value of private land, and this would certainly make it much easier to undertake social development in urban areas.

The more radical members of this group simply dislike the whole idea of anybody making money which they have not actually earned, and it would be difficult to satisfy them, but the more moderate members base their arguments upon:

- The loss of green land or other pleasant environment to the community when development is undertaken for profit.

- The fact that public expenditure on infrastructure (roads, services, transport systems, etc.) has often contributed to the rise in value.
- The way in which decisions taken by the planning authorities are handing windfalls to some lucky landowners and withholding them from others, thus providing strong incentives for undesirable (or even improper) influence upon those decisions.

Several unsuccessful attempts have been made by British governments to deal with the problem, of which the most important were the Town and Country Planning Act of 1947 and the Community Land Act of 1975. However, these Acts tended to inhibit development of any kind, as landowners had little incentive to allow their land to be built on, and preferred to wait until a more right-wing government got into power and repealed the legislation. More recent developments include environmental legislation and new planning guidelines favouring 'brown-field' sites for redevelopment rather than building on hitherto undeveloped land. However, so that the community does not actually lose money over private development, the developer may often be asked to contribute towards infrastructure costs as a condition of the granting of planning approval, and sometimes a developer may make such an offer as part of the planning application. The agreement of these costs is obviously an area where the cost advisers on both sides should be involved. An interesting technique was used in Hong Kong in the 1980s, where the then colonial government was able to act unilaterally and the new underground railway system was largely financed from a tax on the increase in development values in the areas around its suburban stations.

Effect of land values on cost planning techniques

It is possible to try and optimise building costs by looking at such factors as low wall/floor ratios, the avoidance of multi-storey construction, the centralisation of services, and so on, in isolation from land costs. Where land prices are relatively low, or in a public sector organisation where land costs and building costs are not considered together, this may seem to be a perfectly valid way of looking at the problem.

However, where development values and land prices are high the picture alters.

Example

Suppose that the current price of flats in a fashionable urban area is 20% building cost and 80% land cost (and profit) and a piece of land has been bought at a price which assumes that 20 flats can be built on it. If it proves to be possible to build more than 20 flats on the plot, the building cost of these extra flats would be only 20% of their market value, leaving 80% profit.

It would therefore be worthwhile to accept a less efficient building configuration to provide these extra flats. Even if this increased the building cost slightly there would still be a handsome profit. In these circumstances cost planning may develop into an exercise to secure the maximum number of accommodation units on a given site, at the same time optimising the design from the point of view of grants and subsidies (if there are any of these to be had). Unfortunately for the developer, a scheme which squeezes the utmost in hard cash out of the site is rarely acceptable to the planning authorities, who have more regard than the developer for the amenities and appearance of the neighbourhood. Their veto on a scheme is final, apart from the possibility of a lengthy and time-consuming appeal to the central government (which has the same general motivation).

On such projects the costing and income appraisal of many alternative schemes will be required, and this will be the main cost planning contribution. Once a scheme has been approved by the authorities speed may be vital (because of the need to recover a massive investment as soon as possible), and the cost planner will be required to advise on a suitable contractual system to achieve speed with an appropriate measure of cost control.

However, not all private development is of this speculative kind. The client organisation may be wishing to erect a building for their own use – as offices, warehouse, store, etc. In this case the effect of high land costs on the cost planning exercise should be to cause an examination of the economics of possible alternatives. Such alternatives might include carrying out the development in a less expensive area, or possibly changing the client's requirements for a building by solving the problem in another way. An example of this might be to avoid the cost of branch warehouses by a system of daily distribution from the factory. It might be thought that there would be an advantage to a user-developer in building in a high cost area because the firm would always have the value of the site on its books, but there are many disadvantages in having too many of a firm's assets tied up in site values.

Grants, subsidies and taxation concessions

Action under town and country planning legislation is essentially passive, and while it can prevent undesirable development it cannot cause socially desirable development to be undertaken. It can certainly earmark a certain area for a particular type of development, but unless the development itself is to be carried out by a public agency, or unless it is obviously going to be highly profitable, the matter will rest there. In order to make such development attractive to a private investor central or local government may sometimes offer special financial inducements. These are usually offered:

- on a sector basis (e.g. for hotel or other tourist building);
- on a regional basis;
- or both (e.g. for factories in an area of high unemployment).

These incentives will need to be taken into account in preparing budgets for the type of development concerned, and may be of crucial importance in deciding between alternative sites.

The incentives may be of several different kinds:

- *Low rent or rates*. Land (or even completed buildings) may be made available by a public body at a very low rent or with substantial relief from local authority charges, or both. This concession usually runs for a limited period of years, after which more normal conditions will apply, so that some kind of discounted cash flow analysis will probably be necessary.
- *Grant towards capital costs*. The Department of Trade and Industry used to provide grants towards the cost of factory building and plant located in 'areas of expansion', and an increasing number of authorities provide partial funding for development in the fields of tourism, leisure, conservation, etc.
- *Taxation relief*. Taxation relief may be given under some circumstances on development and plant costs. The main disadvantage of taxation relief compared to a grant is that there must be a tax liability against which the relief can be set. It is therefore necessary to make a profit before the benefit can be obtained.

The situation is always changing over the years as different political and economic priorities come into play. A most important factor in assessing the value of a grant is its timing in relation to the developer's expenditure, particularly whether it is a reimbursement or an advance. It should also be noted that other forms of financial encouragement, such as a subsidy paid for each person employed, may need to be taken into account in a development budget, even though strictly speaking it is not a grant towards the development itself. The extent to which government incentives may be nothing more than compensation for straightforward commercial disadvantages, such as high transport costs, will certainly emerge from the cost planner's calculations. It is also worth pointing out that it is possible for government to impose financial disincentives for types of development of which it does not approve. The lack of grants for industrial building in some regions can be seen as such.

A more extreme example of disincentive in post Second World War years was the long-term discouragement of private rental housing development by the control of rents at uneconomically low levels. Although the UK government now wishes to reverse this trend, prospective developers need to be able to look ahead for more than the 5-year term of a government, and the treatment of housing (and other development matters) as a political football by the two main British parties has discouraged a stable property rental market such as exists in other European countries.

Land costs and social development

In the past the costs of site acquisition have usually been ignored in cost planning social development, at any rate as far as the cost planner is concerned. Without any profit a profit-investment appraisal cannot be prepared for such projects, and the cost of the land is just one of the many items on the expenditure side with which no comparison with income can be made. Also there are still many projects where the budgeting stage is bypassed, and the cost planner is concerned with nothing more than meeting a target for building cost which may have been set using some formula or other. However, there is an increasing tendency to use the profit-development type of calculation on any social projects, such as housing, where there is a real or hypothetical income.

Calculating building costs

Although building costs are only one of the ten items making up the total budget, they are still likely in most cases to form the largest single item. However, it must be remembered that at this stage almost nothing is likely to be known about the building except its general size, and therefore it is pointless to go into detail about cost before any designing has been done. This is the place to use one of the traditional 'single price rate' methods of estimating cost introduced in Chapter 1; the size of the building is measured in one form or other and the resulting quantity is multiplied by a single price rate to give the estimated cost.

The cube method

This was the traditional method. All quantity surveyors' offices used to keep a 'cube book'. Whenever a contract was signed the amount of the accepted tender was divided by the cubic content of the building and the resultant cost per cubic foot entered in the book. When an estimate was required for a new project its volume in cubic feet would be calculated and an appropriate rate per cubic foot taken from a previous job in the cube book. The method has largely died out as a result of a number of factors, the main one being that the cost of a building is usually more closely related to its floor area than its cubic capacity.

Even with a 'blunt instrument' such as a cube, it was necessary to have common standards of measurement so that the rate obtained from one project might be fairly compared with another. The RIBA pre-metric set of rules was the one commonly accepted; these involved multiplying the plan area measured over the external walls by the height from the top of the concrete foundation to halfway up the roof if pitched, or to 2'0" (0.60 m) above the roof if flat. Extra allowances were made for projections such as porches, bays, oriels, turrets and fleches, dormers, chimney stacks and lantern lights.

This formula had little to recommend it except the vital matter of uniformity. The additional allowance for a pitched roof (which may well have been cheaper than a flat roof) and the allowance for fluctuations in foundation depth were both very arbitrary.

The Standard Form of Cost Analysis (published by the Building Cost Information Service of the Royal Institution of Chartered Surveyors) still makes provision for cubing but uses a different set of rules.

The superficial area method

This is an alternative single price rate method in which the total floor area is calculated. The convention usually accepted is that the area of the building inside the external walls is taken at each floor level and is measured over partitions, stair wells, etc. The resulting figure is known as the gross internal floor area. In cases where columns to frames are situated inside or beyond the external walls these columns are ignored and the dimensions are still taken from the internal face of perimeter walls. Where there is no external wall at floor level (as in the case of freight sheds) the dimensions are taken from the external faces of columns.

More recent than the cube method, the superficial area method first came into use in the 1940s and 1950s for such projects as schools and local authority housing where the storey heights were reasonably constant. It has the virtue of being closer to the terms in which the client's requirements are expressed, as the accommodation is more likely to be related to floor area than to cubic displacement. Sometimes when a very early estimate is required the only data available may be the approximate floor area, so this method is still widely in use for budgeting as the first stage in a cost planning process.

The unit method

This technique is based on the fact that there is usually a close relationship between the cost of a building and the number of functional units it accommodates. By functional units we mean those factors which express the intended use of the building better than any other. For example it may be:

- number of pupils in a school;
- number of vehicle spaces in a car park;
- number of seats in a theatre;
- number of people housed in a housing development;
- number of beds in a hospital ward.

Several cost yardsticks operated by government departments were based upon this relationship and provided a benchmark of what was reasonable for a given scheme. After all, by costing a single functional unit instead of, say, floor area, the designer has greater flexibility in choosing between the quantity of building and its quality, and the funding organisation has the assurance that it has not paid beyond what is reasonable. We will consider an example of the more general use of the technique.

Example

Suppose a multi-storey car park for 500 cars had recently been constructed for the sum of £3,000,000. The cost per car space would obviously be:

$$\frac{£3,000,000}{500} = £6,000 \text{ per car space}$$

If a similar car park is to be built for 400 cars then it might reasonably be assumed that the cost would be:

$$400 \times £6,000 = £2,400,000$$

all other things being equal.

Single price rate methods generally

Unfortunately all things are not usually equal and consequently a number of judgements must be made about:

- prevailing price levels (due to inflation and market conditions) at the proposed date of tender;
- site difficulties;
- specification changes;
- different circulation and access arrangements, etc.

Once again, as with all single price estimates these adjustments to the analysed figure would be based on the cost adviser's judgement and are critical to success.

Cost of financing the project

Having looked at the role of land costs and building costs in relation to the project it is now time to look at financing costs. This is the intangible component of total building cost; all the other things like buying land, paying a building contractor, paying professional fees, etc., involve paying money to somebody else in exchange for property or services.

As soon as this tangible expenditure starts, however, the developer will be laying out money, and there will be nothing to show for it in terms of income (or use) until the building is ready for occupation or can be disposed of. Hence the cost of lending this money to the development for a period of possibly several years becomes part of the cost of the development itself. When interest rates are high this financing cost can be considerable, as has already been pointed out, so the project should be planned to avoid unnecessarily early expenditure on any part of it.

Where the money comes from

The capital for the development may come from a number of alternative sources.

Sources of finance for private development

- *Bank overdraft.* This is rarely available as a means of financing a whole development, but may be useful for short-term bridging purposes; the money would probably be lent on the security of the development.

- *Loan from merchant bank or insurance company.* This is similar to the last, except that it is possible to obtain longer-term finance from these sources. Interest rates are usually somewhat higher.

- *Capital account.* Where the development is being undertaken for a firm's own use – such as factory, warehouse, etc. – the capital for the development can be regarded as part of the general capital expenditure of the firm.

- *Trading funds.* Property companies will have funds for carrying on their business.

- *Shares in the development.* Speculative development is sometimes financed by paying the builder with shares instead of cash, or by allowing the building firm to develop part of the site for itself. Alternatively a 'joint venture' may be undertaken with another developer.

- *Finance by the intended occupier.* It may be possible to persuade the intended occupier to pay part of the sale price during the erection of the building. This is sometimes the case with houses which are built for sale.

Sources of finance for public (social) development

- *Government funds.* These will be available either as direct finance for government building (e.g. defence works) or as a grant or subsidy for other public sector or social building.

- *Loans from government funds.* These are usually available at a lower rate of interest than money raised on the open market, but are very restricted as to eligibility and amount.

- *Issue of stock.* Local authorities and public boards are empowered to issue fixed-interest loan stock on the Stock Exchange. Full market rates of interest have to be paid, and there is a long-term commitment to these which can prove very expensive if interest rates subsequently fall.

- *Local authority mortgages.* These are a useful way of raising money, as there is a firm commitment by both sides, but for a short period of 3 to 5 years only. Interest rates are usually a little higher than the current yield on gilt-edged securities.

- *Loans from other funds.* A public authority can borrow from its other funds (if it has any) for development purposes.

- *General income.* It is common for a local authority to raise money for development from its charges or council tax – this has the advantage that the money does not have to be paid back!

- *National Lottery.* Grants are available for suitable projects by application to the Arts Council or the Sports Council. Grants are also available at the time of writing from the Millennium fund, although these are unlikely to continue. There is strong competition for money from these sources.

- *Private developer.* In large-scale urban redevelopment a private developer can be required to undertake some social development in exchange for being allowed to pursue a profit-related scheme. Alternatively, a developer can be offered some profitable role within a social development if it contributes to the cost of the public part.

Budgeting for refurbishment work

Before drawing up a budget for refurbishment work, or work involving major alterations to existing premises, a number of difficult decisions have to be made, and it is necessary to consider very carefully how the work will be carried out before establishing even an initial budget. These matters are dealt with in Chapter 22 Cost Planning and Control of Refurbishment and Repair Work, to which reference should be made.

Budgeting for new-build projects on very restricted urban sites

Such projects have many problems in common with refurbishment work, and reference to Chapter 22 would be worthwhile.

Summing-up

The total expenditure on the development will be much higher than the net cost of the building fabric, and will include cost of land, legal costs, demolition or other physical preparation, professional fees, furnishings and fitting out, VAT, cost of leasing or selling, cost of finance and costs of maintenance or disposal.

Building development of any kind requires a plot of land on which it can take place, and which, once used, will no longer be available for any other development, unless the first one is either demolished or converted. The 'development value' of a piece of land is the difference between the market price of the finished development including the land, and the cost of erecting (or converting) buildings on it. Where land costs are very high, cost planning may develop into an exercise to secure the maximum number of accommodation units on a given site, at the same time optimising the design from the point of view of grants and subsidies (if there are any of these to be had).

As soon as tangible expenditure on land purchase, building work, etc., starts, the developer will be laying out money, and there will be nothing to show for it in terms of income (or use) until the building is ready for occupation or can be disposed of. Hence the cost of lending this money to the development for a period of possibly several years becomes part of the cost of the development itself.

Further reading

Laxton, Tweeds. *Laxton's Guide to Budget Estimating*, published annually.

Chapter 9
Some Budgetary Examples

Budget examples

Let us now look at some simple budget examples. These are in no sense 'typical' in that every project has its own particular problems and priorities, but they give an idea of the type of basic calculation which is involved. The method of arriving at the various estimated costs has not been shown, as the concern here is how to manipulate the answers.

The interest rates used for most of the calculations are based on a 'criterion rate of return' – the return which the developer requires in order to make a reasonable profit.

Example

A budget for a profit development for sale (24 town houses in a provincial city)

Costs	£	
Cost of land	1,000,000	
Legal and professional cost of acquiring land, obtaining planning permission, etc.	100,000	
Demolition of existing property	60,000	
Building cost and professional fees in connection	960,000	
Site layout, ditto	40,000	
Agent's charges for selling	20,000	
	2,180,000	2,180,000

Cost of finance (2% per month compound interest – criterion rate of return)

Land, legal costs, demolition – finance on £1,160,000 for 12 months	313,200	
Building, professional fees, and site layout – finance on £1,000,000 for 4 months (av)	80,000	
Agent's fees are paid after sale, no finance charge	—	
	393,200	393,200
		2,573,200

Income

Minimum economic selling price of each house would therefore be £2,573,200 ÷ 24 = £107,210 say £108,000

Anything less than this will not produce the criterion rate of return, and if there is any doubt about obtaining such a figure then the development should not go ahead in this form.

It will be seen that instead of discounting all expenditure and receipts to the commencing date of the scheme, they have been carried forward to the end of the scheme and compound interest added. This has the same effect – it is easier to do it this way in this instance as it is the future selling price which we are interested in and not its present-day value.

A further example

A budget for profit development (for rental offices)

	£	
Costs		
Cost of land	500,000	
Legal and professional cost of acquiring land, obtaining planning permission, compensation, etc.	170,000	
Demolition	60,000	
Building cost and professional fees in connection	700,000	
	1,430,000	1,430,000

Cost of finance (2% per month compound interest – criterion rate of return)		
Land, legal costs – finance on £670,000 for 36 months	696,800	
Demolition – finance on £60,000 for 18 months	25,800	
Building, professional fees, and site layout – finance on £700,000 for 10 months (av)	154,000	
	876,600	876,600
Total capital cost at completion		2,306,600
	say	2,310,000

	£	
Income		
Annual income from rents (figure given by valuer)	600,000	600,000

Less		
Allowance for vacant tenancies (voids)	50,000	
Maintenance, repairs and redecorations (excluding tenants' responsibilities)	30,000	
Cleaning, heating, lighting and council tax on public part of building (staircases, entrance halls)	50,000	
Management expenses, including caretaker, arranging lettings, and collecting rents	18,000	
Sinking fund to replace capital at end of 40 years at $2\frac{1}{2}$% pa on £2,310,000 less cost of land (which will still be there when the building is demolished) £1,810,000 at 1.5p	27,150	
	175,150	175,150
Net annual income		424,850

This gives a return of 18.4% on the capital of £2,310,000.

Alternative format for last budget

An alternative way of setting out this last budget would be to start with the estimated income as the criterion. This is the approach a developer would be likely to favour.

Income	£
Estimated annual income from rents	600,000
Less expenses (excluding sinking fund)	148,000
	452,000

Note: The cost of the sinking fund cannot be accurately assessed at this stage, as the capital value of the buildings, etc., is not known. It could be approximated thus:

Capitalised value of income at $18\frac{1}{2}$% is approximately £2,443,000. Say 25% of this represents residual land value, which will not need to be replaced.

	£
Sinking fund to replace £1,832,000 at end of 40 years at $2\frac{1}{2}$% (at 1.5p)	27,483
Estimated net annual income	424,517
Net value of 40 years' annual rent £424,517 at 18% (criterion rate of return) at £5.40	2,292,400

This represents the capitalised value of income at the date of completion in 36 months time. Discounted to present value at 2% per month at 49.0p, present value.of income = £1,123,270

Costs	£
Estimated legal and other costs of acquiring land	170,000
Estimated cost of demolition of existing buildings (PV of £60,000 in 18 months time at 2% per month = 70.0p per £)	42,000
Estimated cost of building and professional fees (PV of £700,000 in (average) 26 months time at 2% per month = 59.8p per £)	418,600
	630,600

The maximum amount which can be paid for the land at the present time to give the required rate of return would therefore be £1,123,270 (discounted present value of income) minus £630,600 (discounted present value of expenses) = £492,670 (say £500,000). At this stage the calculations would have to be based on the minimum amount of accommodation for which the client could expect to get planning permission, unless an outline scheme had already been approved.

It should be noted that taxation provisions, which are constantly changing, have been ignored in these calculations but cannot be ignored in practice.

Sensitivity analysis

Budgets are usually prepared today by using computers. In these circumstances it is possible to examine the effect of changing the values of the variables either singly or in combination. This will enable those variables which are crucial to the calculations to be identified. Very often in profit development the overall time for designing and building the project is of supreme importance, and simulations might well be carried out to assess the effect on construction costs and financing charges of different time scales and fast-track methods of procurement. Certainly it will be seen in the above examples what a large proportion of the total cost is represented by financing charges. It would be easy to reduce the length of the programme on paper, the more important matter is whether this would be achievable in practice, allowing for all the things that can go wrong.

Need for caution

Schemes founded on over-optimism tend not to have a happy outcome. In fact, some developers require that budgets be prepared on a 'mid-to-worst-case' scenario, knowing that construction costs tend to escalate but funding has to be obtained 'up-front'. In a recent case the developer actually told the cost planner to presume a figure for construction costs which was likely to reduce during the design development and construction stages, and which would not be exceeded without very good reason.

Stage 2
Designing to the Budget

Chapter 10
Design Method and Cost Advice

When the budget has been determined the process of design can begin.

The building design process

The building design process is a complex interaction of skills, judgement, knowledge, information and time, which has as its objective the satisfaction of the client's demands for shelter, within the overall needs of society. Ultimate satisfaction is obtained when the 'best' solution has been discovered within the constraints imposed by such factors as:

- statutory obligations;
- technical feasibility;
- environmental standards;
- site conditions;
- cost.

Problems arise, however, in establishing which of the alternative options available is the 'best', as some factors such as personal comfort or visual delight are difficult to measure. Neither is it easy to translate all attributes into a common unit of value, for example excessive noise levels compared with higher initial cost. In practice, compromises in the client's demands are nearly always necessary to keep within the constraints. The role of the design economist/cost planner is to provide information with regard to initial and future costs so that the design team can make decisions knowing the cost implications of those decisions. It is not usually the building economist's responsibility actually to provide 'value' as this must be the province of the team as a whole – of which, however, the building economist should be a contributing member. In theory the team will pool their combined knowledge for the benefit of the client, whose representative should, wherever possible, be a member of the team and play a part in the corporate decision-making process. As with any group activity the composition of the team should be carefully planned to avoid vociferous members unduly influencing the final decisions.

A design economist who is to be an effective member of the team:

- must have a clear idea of when the major decisions of cost significance are to be made, so that information can be provided at the crucial time;

- must have the techniques, knowledge and experience to provide answers to questions of cost that will be posed as the design is refined;
- must understand the manner in which the design team, and in particular the architect, thinks and operates – commonly referred to as the design method.

Although each individual designer adopts a different approach, as there are many ways of solving the same problem, it is possible to identify some of the common techniques which are very often incorporated in a typical approach to the task of achieving good design. This knowledge can be used to select the form of cost advice most appropriate to a particular problem.

The design team

It is not possible to place all the responsibility for establishing a successful cost solution solely on the shoulders of the building economist/cost planner or quantity surveyor. The architect must cooperate in, and contribute to, the cost planning process, cost being one of the several constraints that have to be faced in varying degrees, depending on the type of project and the client's financial resources. In this respect the architect needs to have a reasonable grasp of the factors affecting the cost of the project and the options that are available within that constraint. This knowledge will often be gained by experience, especially if the firm specialises in one particular type of contract, e.g. housing, factories, hospitals, etc. Even without this experience the architect should be aware of the very basic design and cost relationships shown below.

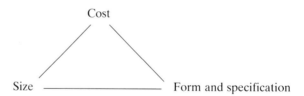

The advantage of illustrating this triangular set of relationships is that any two of the factors can be seen as functions of the remaining one. For example, if the size of the building is fixed, together with the form and specification, then a certain cost will be generated. Conversely, if the cost of the building is established together with its size (as is the case with some government yardsticks), then this constrains the form and specification that can be chosen. Alternatively, if the shape and quality standard of the specification is declared together with an established sum of money, then the amount of accommodation is the design variable which is limited. Since one factor must be the resultant, it is never possible to declare all three in an initial brief.

This is an over-simplistic view of the cost system but it is a starting point in the understanding of the complex relationships which exist between design and cost. It is the skill of the design team in achieving the right balance between these factors that makes a project a success or a failure.

The RIBA Plan of Work

Recognition for this team approach to design is included in what has come to be known as the *RIBA Plan of Work*. This is a model procedure dealing with some basic steps in decision-making for a medium-sized project, and is included in the *RIBA Handbook of Architectural Practice and Management* (Volume 2). In this procedure the responsibilities of each member of the design team at each stage of the design process are identified. A design-tender-construct procedure is envisaged and from the quantity surveyor's point of view it anticipates and incorporates some of the cost planning techniques described in later chapters. The pre-tender procedure includes the following stages:

Stage A	Inception	Appointment of design team and general approach defined.
Stage B	Feasibility	Testing to see whether client's requirements can be met in terms of planning, accommodation, costs, etc.
Stage C	Outline proposals	General approach identified together with critical dimensions, main space locations and uses.
Stage D	Scheme design	Basic form determined and cost plan (budget) determined.
Stage E	Detail design	Design developed to the point where detailing is complete and the building 'works'.
		Cost checking carried out against budget.
Stage F	Production information	Working drawings prepared for tender documents.
Stage G	Bills of quantities	

It is of course recognised that a certain amount of overlap is bound to occur in practice. In recent years the RIBA Plan of Work has come under fire from a number of quarters because of the inherent inflexibility of its procedures, and because it tends to delay the letting of the contract which in times of rapid inflation can severely penalise some types of client. However, it was never intended as a rigid set of rules and the procedures adopted for any contract should be those which best suit the prevailing economic climate and the particular interests of the client. The plan has been included at this point in order to give an overview of the complete design process, but the emphasis in the rest of the chapter will be on the very early design period, prior to sketch design, when the critical decisions affecting cost are usually taken.

Comparison of design method and scientific method

To understand the concept of design it is useful to compare design method with the traditional approach of scientific method as shown in Fig 10.1. To establish a scientific law:

- An observation is made in nature and an inference drawn, e.g. light passing through a prism breaks down into several colours.

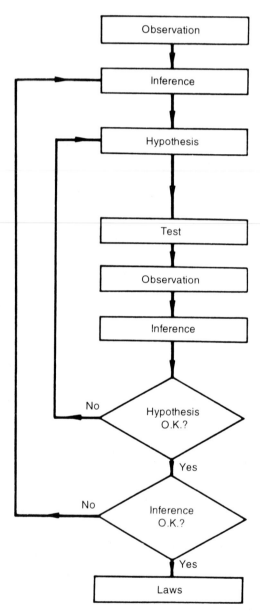

Fig. 10.1 Scientific method.

- A hypothesis is set up (e.g. white light is a combination of several colours).
- Tests are applied to establish whether the hypothesis is true.
- If the results of the tests conclude that the hypothesis and original inference are correct then a scientific law can be established based on the hypothesis.
- If not, a new solution must be set up and tested until the tests corroborate the hypothesis.

Design can be said to follow a similar pattern:

- The brief is observed and some inference obtained.
- A hypothesis is set up in the form of a model (e.g. a drawing).
- This is tested by evaluation to see whether it 'works'.
- A check is made to see whether it complies with the interpretation of the brief, and if it does then the design is accepted.

There are, however, some very important differences. The design team cannot 'loop' round the system producing new hypotheses (i.e. designs) *ad infinitum*. They work within the constraints of time and cost, and they must produce a final solution within the design period set by their client and within the fee structure by which they are paid. In addition, much of their creative work can only be evaluated in terms of social responses (such as aesthetic appeal), which at the present time do not have a satisfactory quantitative measure by which they can be tested.

Consequently, the team's objective tends to change from – 'What is the best solution for our client?' to 'How can we produce a design which satisfies as many of the demands of the brief as possible in the time available?'. To arrive at a satisfactory solution, some kind of strategy, very often incorporating 'rules of thumb' learned from previous experience, is used to narrow down on a particular range of alternatives. The responsibility of the design economist/cost planner in this search for a satisfactory solution should be to indicate to the team where it should look in terms of form, quality, spatial standards, etc., to solve the cost problem. This will avoid wasting time on abortive designs which will not meet the cost criteria, and will thereby increase the chance of arriving at the 'best' solution in the time available. In other words, the building economist should contribute to the overall design strategy.

A conceptual design model

A good deal of research work has been undertaken to establish a general pattern for design. The Building Performance Research Unit at Strathclyde University has put forward the following view, which is shared by several other writers. They suggest that design consists of three stages as follows:

(1) *Analysis*. Where the problem is researched in order to obtain an understanding of what is required.
(2) *Synthesis*. Where the information obtained in analysis is used to converge on a solution at the level being investigated.

(3) *Appraisal.* Where the solution is represented in some form which is then measured and evaluated.

Quantity surveyors have traditionally been involved almost exclusively in the latter part of the 'appraisal' process, and yet the real decision-making role is in the analysis and synthesis stages. New techniques are therefore required to enable the design economist to contribute to understanding and solving the design problem. It was assumed by the Strathclyde team that decision-making became more detailed and refined as time went on (from, say, concepts of building form and spatial arrangement through to the eventual choice of iron-mongery) and that at each stage different methods of problem solving were adopted. Any cost technique must therefore follow a similar pattern, which may involve 'coarse' measurement and evaluation in the very early stages and a more reliable measurement and cost application when more information becomes available. It is with these thoughts in mind that cost models have been constructed which can input information at each level of refinement.

One of the criticisms that can be made of traditional cost planning methods is that they do very little to contribute to the pre-sketch design dialogue, where all the major decisions of form and quality tend to be taken. Current research suggests that there is a heavy commitment of cost prior to a sketch design being formalised. This may amount to over 80% of the final potential building cost, leaving perhaps only 20% available actually to be 'controlled'. Figure 10.2 illustrates the point in a hypothetical diagrammatic form. With an improved understanding of design method it may be possible to input information prior to sketch design, which will reduce the number of abortive design solutions needing to be produced.

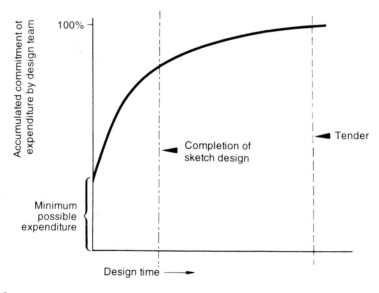

Fig. 10.2

Design techniques

It is not possible to describe fully the vast range of techniques applicable to design in a chapter or book of this nature. Those interested in developing their understanding in this area of study should consult other publications (see the reading list at the end of this chapter). It is possible to identify some decisions which form the basis of most design approaches. In very simple terms these can be illustrated by some general questions relating to matters of principle affecting the design. These include the following:

'What are the constraints within which I have to work?'

The constraints on a project can be of three kinds:

- *Constraints imposed by physical factors.* These very often relate to the site, and include the position of boundaries and easements, the method of access, the nearness to service supplies, any visual aspects and views, soil and environmental conditions and adjacent structures. In addition to the site, the physical performance of materials acts as a limitation and may exclude the use of a particular specification. Constraints imposed by physical factors are usually fixed and therefore provide very clear boundaries to the design problem.

- *Constraints imposed by external bodies.* These, however, can often be the subject of negotiation. Planning requirements are usually the result of the policy of the local planning committee and are often open to interpretation – hence the facility for appeal when these interpretations conflict. Some building regulations can be waived if the appropriate authority can be convinced that an alternative construction form or arrangement is satisfactory. Resolving these problems can be a time-consuming business, and until they are settled the solution to the design problem has to remain flexible. However, once they are defined they again provide a clear boundary to the design problem.

- *Constraints imposed by the client body and its advisers.* These tend to be far less stringent than the previous two categories, and can sometimes be compromised more easily. Even here some may be inflexible due to a specific demand which takes precedence over all other needs. An example may be the requirement to design a form which envelopes an expensive manufacturing process; because the plant is so expensive the form of the building must take second place. Another example is a cost limit which is subject to the financial standing of the client; this may however be imposed by an outside body such as a government department or finance company who will define the cost that it sees as being realistic. Most self-imposed constraints are able to be reconsidered in the light of experience and gradual evolution of the design solution.

Definition of the constraints is of enormous assistance in containing the design solution. They help in narrowing down the range of possible solutions, which are for practical purposes almost infinite without them. It is the responsibility of the design economist to account for these controlling factors in the budget and, therefore, set a realistic strategy of cost. Ignorance of these issues will possibly result in abortive effort and a less than satisfactory service to the client.

'What are the priorities of the scheme?'

In a sense this is the question that sets the self-imposed constraints and whose solution should provide value for money. If priorities can be ranked and given their due importance in solving the design problem then it should be possible to spend the client's money in accordance with these requirements. This would be the ideal design and would provide the optimum solution. Unfortunately 'ranking all the priorities' is an extremely difficult task. For example, should maintenance-free windows take priority over an improved reduction in noise levels between rooms in a commercial office block? It is difficult to compare the two, and it is even more difficult to award a satisfactory weighting. However, this kind of decision is at the root of good cost budgeting and, if possible, it should be made in conjunction with the designer. The theoretical aim should be to spend money in the same order and importance as the priorities.

Whether this can ever be achieved is a debatable point, as an item which is not given a high priority by the client may yet be essential, for example easy-to-clean windows on a multi-storey block. To provide this item may be more expensive than providing an item of greater ranked importance. This does not mean that the higher ranked item is of less value to the client, but just that because of external economic forces and the nature of buildings the client has to pay more to achieve the desired objective. This emphasises the difference between the two concepts of 'value' put forward by the economist Adam Smith: value in exchange or value in use. In most buildings of a public or social nature it is value in use which is being considered, whereas in a speculative housing or commercial development it is value in exchange (which is much more dependent on scarcity factors) that is the major concern. There is, of course, a link between the two because the greater the degree of user satisfaction, the more likely it is that the client will be prepared to pay more for his building. Value for money is achieved when the priorities of the client, for example profit, symbolism, welfare, religious worship, etc., have been successfully balanced with the initial and future costs allowable for the development.

'How much space is required?'

The purpose of building is usually to provide space for a particular activity within which the climate is modified so that the activity may function more efficiently. Other considerations such as the environmental and social impact of the building and its activity arise out of this prime need. It is therefore usual for the client's brief to give an indication of the usable area required, but even where this is stated there is considerable flexibility allowed, for example the circulation area (corridors, waiting areas, lifts, public areas, etc.) is not usually included in the list. In addition, the multiple use of space (e.g. assembly halls doubling up as dining halls in schools) may allow a more efficient use of a certain area and reduce the overall requirement. Part of the design problem is to discover the most efficient use of the spaces required to satisfy the client's brief and to arrange the areas in such a way that circulation is kept to a minimum.

As there is a strong correlation between the area of a building and its cost, the design economist should be an active participant in the discussion of areas and their spatial arrangement. Statutory requirements with regard to means of escape in cases of fire, disabled access and disability requirements, health, and hygiene, etc., in conjunction with the client's requirements for efficient movement in the building, will provide an indication of the minimum areas allowable for circulation and ancillary purposes. A careful study of these needs will provide the constraints for area and thereby indicate the balance between quantity and quality of building that can be achieved within a given cost limit.

'What arrangement of space is required?'

This question is heavily influenced by the amount of space needed. In answering it, a good designer's knowledge can be exercised to enormous advantage. Indeed, determining spatial organisation has been recognised as one of the most fundamental of the architect's range of skills.

Despite the advent of computer programming techniques which attempt to optimise the positioning of space, the ability of the human architect to take into account a large number of factors and bring them into a suitable relationship has not yet been surpassed.

A number of techniques have been developed to assist the architect in this important task, and perhaps the most common is the association matrix. In this design aid the relationship between spaces is identified in a table rather like a 'mileage between towns' indicator in a road atlas. The figures in the table, however, will relate to the 'cost of communication or movement' related to a unit of distance between spaces rather than measured distance. For example, the salary cost of the managing director spending time in walking one metre may be five times that of the administrative clerk. It follows that the MD's office should be given priority in being closer to the centre of communication.

Problems arise in this technique when the subject has no direct wages cost, for example the casualty patient in a hospital who does not cost the hospital administration any salary. Without an artificial weighting factor, hospitals might be designed with the patient always taking the longest route!

The simplest form of the matrix is that in which a simple weighting system is used to define the ease of movement between spaces. To illustrate the use of such a table let us take the example of a new administration/accommodation wing added to the existing appliance garage of a fire station. The spaces required by the fire authority are as follows:

BLANKSTOWN: EXTENSION TO EXISTING FIRE STATION

Space	Area (m^2)
(1) Administration area including watchroom	120
(2) Lecture room	80
(3) Firefighters' dormitory	140
(4) Television room	25
(5) Recreation room	150
(6) Mess and officers' lounge	200
(7) Visitors'/administration toilet and wash room (male and female)	25
(8) Visitors' entrance	10
(9) Firefighters' entrance	20
(10) Firefighters' ablutions area	40
(11) Appliance garage (existing)	—
	820

The first step is to identify a suitable scale for the ease of movement required between spaces. The following may be considered reasonable:

Weighting	Degree of 'ease of movement'
5	Essential
4	Desirable
3	Tolerable
2	Undesirable
1	Intolerable

By considering the relationship of each space to the other spaces an association matrix can be established using the above key.

It can be seen from the developed matrix (Fig. 10.3) that the crucial space is the existing appliance garage. This is fairly obvious as the efficiency of the station is centred around the ability of the firefighters to reach their equipment quickly when an emergency call is received. The next step is to arrange the spaces in a diagrammatic form to show the desirable 'clusters' of accommodation. This is usually shown by means of a 'bubble' diagram identifying the spaces and their required links. In arranging the groups of spaces, consideration needs to be given to important constraints, such as the site, which may not allow ideal groupings to be made. For example, the parking and practice area required behind the fire station may restrict the usable site area and force a two-storey solution. The resultant bubble diagram incorporating the site constraint and association table may be as indicated in Fig. 10.4.

The bubble diagrams reinforce the relationship between spaces and emphasise the view that the appliance room is the controlling factor in any arrangement. In addition, the size of the spaces is illustrated, and a visual indication of the likely room clusters is given ready for incorporation into a suitable building form. The efficiency of spatial arrangements in building is not a field that has involved the cost adviser to any great extent in the past, apart from commenting on the ratio of circulation space to usable floor area. In buildings in which ease of movement

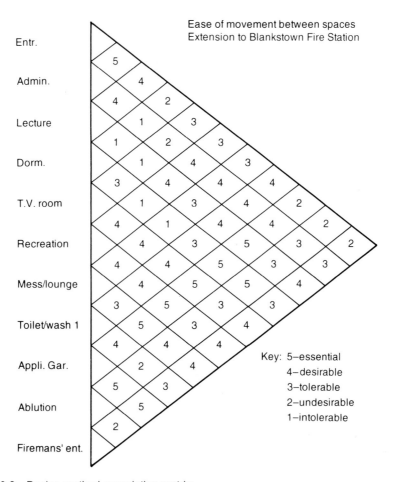

Ease of movement between spaces
Extension to Blankstown Fire Station

Entr.

Admin.

Lecture

Dorm.

T.V. room

Recreation

Mess/lounge

Toilet/wash 1

Appli. Gar.

Ablution

Firemans' ent.

Key: 5–essential
4–desirable
3–tolerable
2–undesirable
1–intolerable

Fig. 10.3   Design method association matrix.

between spaces is a priority, such as those housing industrial processes, this area of study may prove fruitful for cost research and investigation as the layout has a major effect on the overall economy of the scheme.

'What form should the building take?'

Answering this question involves another of the essential design skills of the architect, who should be able to translate the functional spatial arrangement of the bubble diagram into a building form that will reflect the relationships determined. In doing so the architect will also be aware of the constraints which usually reduce the design options considerably, the largest constraint in many cases being the site itself. In other cases, planning requirements or cost limits may be paramount.

In the case of the Blankstown Fire Station the available site area is considerably reduced by the need for a practice yard at the rear. The orientation

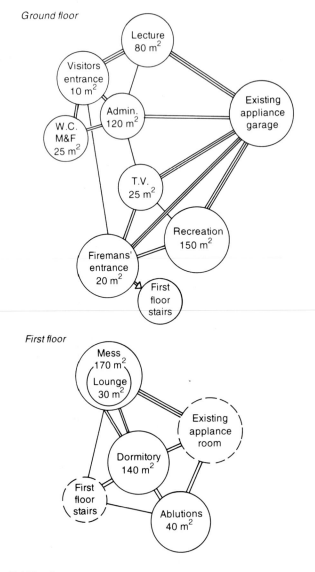

Extensions to Blackstown Fire Station
Note: Numbers of interconnecting lines shows strength of association

Ground floor

First floor

Fig. 10.4 Bubble diagram.

is also fairly limited because of the desirability of having a frontage on to the High Street. The resultant plan is an attempt to segregate administrative areas from recreation/work areas and at the same time provide speed of access to the appliances from any part of the building or surrounding areas.

Figure 10.5 illustrates the first sketch of a design that may be suitable. Note that from force of circumstances the architect has had to use open space in rooms for through access to the appliance garage. The visitors'/administration toilet

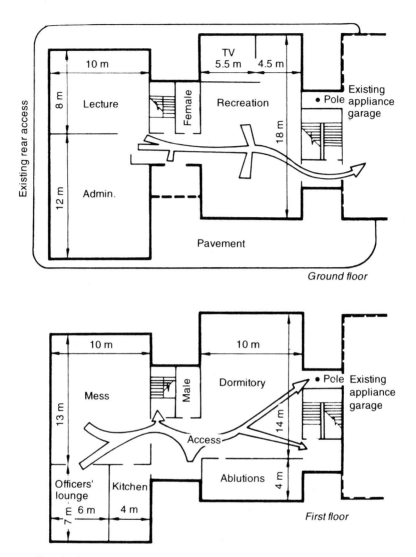

Fig. 10.5 Sketch plans.

facilities have had to be split between ground and first floor to obtain male and female facilities. In addition an attempt has been made to isolate noisy areas from the quiet/rest rooms, and the recreation block has been set back, thus providing passing motorists with a better view of the garage doors – very necessary in an emergency.

It is interesting to note that it is at this stage that the cost adviser is usually first asked for an estimate, yet it is clear that there is already a heavy commitment to space, circulation and form standards. In addition, the specification standard will probably have been indicated in the brief and therefore the degree of cost control that can be exercised from this point on is severely limited. This reinforces the view that cost advice is most effective if given prior to the sketch design.

A request to design a more compact form once the architect has already produced a solution is likely to involve considerable effort – and also resentment. One of the aims of good cost budgeting must be to avoid abortive effort and therefore to advise where cost-acceptable solutions are likely to be found.

'What is the level of specification?'

This is very often the decision that has the least external constraint. It is also the design decision that suffers most when cost reduction is required at a late stage of the design process.

At the present time there are very few numerical techniques that attempt to measure quality. It is therefore a decision that is based largely on intuitive judgement arising out of the previous experience of the design team.

The client's brief will have attempted to give an indication of need, perhaps in the form of a requirement for soft or hard finish in certain rooms, maintenance-free exterior, and so on. Standards of environmental comfort and the need for prestigious public areas, etc., will also have been conveyed by the client.

In these circumstances the selection by the designer has to be narrowed to those materials that will fulfil the function required, and a selection made based on comparative performance. An important aspect of that performance will be the economic considerations of installation and durability that will generate capital, maintenance and operational costs.

One role of the cost adviser should be the production of alternative estimates for each major solution, in order that the cost implications of a decision can be known.

Generally

The above series of questions is a gross oversimplification of the design problem and the text has been orientated towards those factors which have cost implications. The object has been to give the student, in disciplines other than architecture, a grasp of the early stages of the design problem and the potential for economic advice, and in particular to dispel the common misapprehension that the architect starts work by sketching a building form neatly divided up into rectangular rooms and passages.

However, it should not be assumed that a linear chronological order has necessarily been implied by the sequence of the above questions. In some cases a linear sequence may be the method adopted for the solving of a particular problem, but in the vast majority of cases there will be a good deal of overlap. There may be simultaneous decisions covering all the points discussed, and possibly reversal of the process.

The above procedure has been largely a process of designing from the 'inside-out'; from internal space requirements to external form. In buildings where external form is important or the site constraints are very tight then it may be necessary to design from the 'outside-in'.

Each individual designer's methods and techniques will have been developed to suit that particular person's approach – not every architect, for example, will use numerical methods to assist in deciding on circulation priorities.

There are, of course, a large number of design decisions still to be made on Blankstown Fire Station and these include:

- the working and arrangement of the building services;
- the sizing and choice of the structure;
- the fixing of components;
- the detailing of the specification;

and so on. Some of these decisions will be made in conjunction with those already outlined, and in fact may depend upon those decisions and vice versa.

Design cannot be neatly contained in water-tight compartments of sequential decisions, and this does make the input of information by other consultants more difficult.

An understanding of design method will, however, help the cost adviser to know when cost advice is likely to be most helpful in its contribution to solving the design problem. It will also assist in determining the degree of refinement in cost advice that will match the stage of detail in the design process.

Recognition of design methods in cost information systems

It is generally recognised that any cost information system must be compatible with design method. However, unless cost is of paramount importance it should not dictate the approach the designer takes – if the financial tail wags the project dog this will not normally be conducive to satisfying all the client's requirements.

The cost adviser should also be aware that striving to achieve optimisation in cost is only a part of the total objective. While all consultants would like to achieve optimisation in their own subsystem of activity, it is the performance of all systems acting together that will determine the degree of client satisfaction.

Two examples will suffice to demonstrate recent attempts that have been made to marry cost advice to design method in the pre-sketch design stage, when the major cost-significant decisions are made. The systems are not fully developed but they give an indication of current trends.

Since these are the first of many cost control systems at different levels which we shall be encountering it is necessary to make a point which will be repeated on appropriate occasions. Systems such as the following represent overall strategies within which a cost control process operates.

It is very rare in practice, however, for the cost adviser to have sufficient resources available to be able to carry out every single step of the strategy in respect of every aspect of the design, and it is doubtful that this would be a very economical thing to do in any case.

Just as a general commanding an army cannot afford to attack all along a front but must concentrate its efforts at the most sensitive points, so the cost adviser's resources must be used where they will do most good. Making a wise decision on this issue is probably one of the most important and difficult parts of the cost adviser's job, and is one of the things which makes it a truly professional task.

An evaluative system

In this system the cost adviser attempts to evaluate the cost implications of design knowledge and decisions immediately they are postulated. Design is considered to be the arrangement of spaces into a building format. The procedure could be as follows:

(1) Designer receives brief with room sizes given.

Cost adviser evaluates cost of alternative finishes and room contents including service outlets.

(2) Designer organises rooms into groups and location in a building shape and form.

The form generates a type of production method and site transportation which the cost adviser evaluates.

(3) Designer looks at the structure needed to support the chosen form.

Cost adviser evaluates cost of alternative forms of support.

(4) Designer envelopes the structure.

Cost adviser evaluates alternative specification of external walls and roof.

(5) Designer organises circulation and service runs, etc.

Cost adviser evaluates alternative arrangements of lifts, service runs and corridor space.

There are two problems with this method.

- It assumes a 'linear' design method, with each event occurring after the other, although in practice these decisions are very often taken simultaneously
- It depends on the generation of information by the designer before evaluation can take place, and therefore gives no indication as to where to look for a good cost solution prior to a committed decision being made on building form.

It is, however, a significant step forward in providing an evaluation of the decision-making process prior to sketch design.

A strategic cost information system

By use of this method the cost adviser attempts to explore a range of possible solutions available to the designer using a structured 'search' process in which the designer is involved. The object is to identify a cost 'strategy' which the designer can employ in the synthesis of building form and which will avoid abortive redesign at a later date. The procedure may take the following form:

(1) Design team receives brief with accommodation area and quality standard given.

Cost adviser explores the cost of different components of the building according to changes in the major design variables (e.g. the area and shape of a bay in a structural frame).

(2) Design team selects the specification and parameters for each component from the explorations undertaken.

Cost adviser explores the use of these components and parameters in buildings of different shape and height according to a predetermined series of building descriptions, e.g. plan shape, number of storeys, density of partitions, etc. A cost table is produced of the feasible solutions available within the identifiable constraints.

(3) Design team identifies the lowest total cost from the table and uses it as the point of reference for selecting any other alternative.

Cost adviser contributes to the discussion of an alternative solution.

(4) A solution is chosen from the table by the design team.

Cost adviser provides a breakdown of the major component costs which is then adopted as the point of reference for the initial cost plan for use in traditional budgeting. The design variables incorporated in the selected solution are communicated to the designer.

(5) Designer uses the parameters and descriptions as guidelines in the preparation of the design solution.

Cost adviser uses traditional budgetary procedures to maintain control.

The descriptions conveyed to the architect for the design strategy should not be so rigid that a 'strait-jacket' is imposed resulting in only one possible solution. Rather, they should be 'coarse' measures which still allow reasonable scope for using the designer's skills of modelling and spatial organisation within the strategy adopted by the team as a whole. The choice of descriptors is important in making sure that the creativity of the designer is not stifled.

The advantage of this method is that it attempts to undertake a systematic search through possible solutions in order that the design team may know where the better cost propositions are likely to be. It identifies a 'least cost' solution which can then form the point of reference for any alternative selection. It is, therefore, possible to obtain a gauge of value by comparing a specific choice with the lowest cost.

There are, however, quite a number of problems. The setting up of 'models' that will cope with the large number of components and design variables is an onerous task, even though the models need not be very refined at this stage of the design process. It is also extremely difficult to produce a satisfactory range of descriptors which will be simple enough for the design team to incorporate into their thinking and also comprehensive enough to allow a reasonable evaluation of the building as a whole.

Both the evaluative system and the strategic approach depend heavily on the use of computers to provide the information sufficiently quickly for decision-making purposes.

Computer usage

As outlined in Chapter 2 the development of computer technology has meant that machines are available, at less than the cost of one man-month, that will cope with these types of problems. The real expertise is in formatting the system to suit a particular type of building, and quantity surveyors will need access to this expertise if they are to provide this improved service to the building client. Without the ability to re-use computer data and systems most cost advice is likely to be too expensive to be provided on any but the largest projects.

Summing-up

An understanding of design method enables the cost adviser to know:

- the time when cost advice will be most effective;
- the type of advice that needs to be given;
- the objectives of the design team additional to minimisation of cost.

It would appear that the earlier the advice is given, the greater the chance of the advice being incorporated into the final design solution. The need is for techniques which contribute to the analytical understanding of the problem and assist in the convergence on the best solution in the shortest possible time. The development of cost models and computer techniques seems to show a possible way forward in achieving these objectives.

Further reading

Broadbent, G.H. (1995) *Emerging Concepts in Urban Space Design*. Spon, London.
RIBA (1998) *Handbook of Architectural Practice and Management*, Vol 2. RIBA, London.

Chapter 11
Introduction to Cost Modelling

Prototypes

When an industrial firm is about to manufacture a new product they first of all build a prototype. They do this for a number of reasons, including:

- To identify and solve three-dimensional problems that were not apparent in the drawings.
- To identify the tools required for production.
- To help estimate the cost of production.
- To test and evaluate its functional performance.
- To test its marketability.
- To provide a sample for a customer as evidence of the quality standard to be achieved.

By building, manipulating and testing the prototype the manufacturer can iron out and avoid future problems when the product is actually manufactured, supplied or sold. However, it is not always feasible to build a prototype, and in the case of buildings there are particular problems:

- Buildings are very large.
- They are intended to last for a very long time, and it would be difficult to simulate this in prototype testing.
- Normally only one production model is to be built and the prototype costs cannot be written off over a large production run.

Therefore, because of the time needed for construction and the great expense and the individual nature of buildings, it is just not realistic to construct trial examples of a whole project to see whether or not it will work satisfactorily.

It may, however, be possible to build and test samples of major components, and this possibility may need to be taken into account in cost planning a building which is to incorporate some measure of innovation or a great degree of repetition.

It might be interesting at this point to mention the Sydney Opera House, a building of great prestige and enormous innovation, where the construction and testing of prototypes and mock-ups was undertaken on an exceptional scale. The result of this was that very few problems were encountered in the use and maintenance of the building, considering the potential for trouble in a project of this nature, but of course a considerable penalty was paid in terms of construction time and cost.

Other types of model

If physical prototypes are not normally possible the design of a building must still be assessed in some way to try and ensure that the demands of the client are going to be satisfied. For the sake of expediency and cost it is, therefore, necessary to construct models that represent the real situation in another form, or to a smaller scale, so that a realistic appraisal of performance can be made. These models can be:

- physical (as in a three-dimensional architect's balsa model);
- three-dimensional computer graphic 'walk-throughs' and 'fly-bys';
- mathematical (as in a heat loss equation)
- statistical (where some collected information indicates a certain trend).

Cost is one of the measures of function and performance of a building and should therefore be capable of being 'modelled' in order that a design can be evaluated. In recent years a considerable effort has been made to construct models that will help in the understanding and prediction of the cost effect of changing design variables.

Cost modelling may be defined as the symbolic representation of a system, expressing the content of that system in terms of the factors which influence its cost. In other words, the model attempts to represent the significant cost items of a cash flow, building or component in a form which will allow analysis and prediction of cost to be undertaken. Such a model must allow for the evaluation of changes in such factors as the design variables, construction methods, timing of events, etc.

The idea is to simulate a current or future situation in such a way that the solutions posed in the simulation will generate results which may be analysed and used in the decision-making process of design. In terms of quantity surveying practice this usually means estimating the cost of a building design at an early stage to establish its feasibility.

Objectives of modelling

Models for estimating and planning costs have evolved gradually. Their adoption by the profession at large has led to the establishment of what might be called 'traditional' techniques. These will be discussed and referred to in this chapter, together with some that have been suggested but have not yet been widely practised.

At this stage it is useful to consider what a good cost model should be attempting to achieve. Traditional and future models can then be tested to see to what degree they comply with this set of requirements. The broad objectives can be listed as follows:

- To give confidence to the client with regard to the expected cost of the project, i.e. economic assurance.
- To allow the quick development of a representation of the building in such a way that its cost can be tested and analysed.
- To establish a system for advising the designer on cost that is compatible with the process of building up the design. This should be usable as soon as the designer makes the first decision that can be quantified, and should be capable of refinement, to deal with the more detailed decisions that follow thereafter.
- To establish a link between the cost control of design and the manner in which costs are generated and controlled on site. This involves dealing with the cost of resources at as early a stage as possible to aid communication between the design team and those responsible for managing the construction process.

Linked to these objectives will be some guiding principles which can be applied to the way we approach and verify the model. These may be summarised as follows:

- The degree of refinement of the model should be tailored to suit the stage of design refinement, and should call on as much design information as is available at the point in time when the model will be used. The cost data applied to this information should represent the degree of reliability that can reasonably be expected from an estimate at that stage.
- The cost data in the model should be capable of updating and evolution in the light of changes in external market and environmental conditions, without too much time being involved.
- The representation of the building or component in the model should bear an understandable relationship to design method (e.g. the arrangement and use of space) and if possible the manner in which the costs are incurred (e.g. the production method). Ideally the model should show the relationship between the client's objectives expressed in the brief and the subsequent cost of resources used to achieve these objectives. This is, however, an extremely difficult task as many of these objectives will be of an intangible nature.
- The model should cope with constraints imposed on design and be able to test the feasibility of a proposed solution within these constraints in order that definite decisions can be made.
- The results of the model should enable this knowledge to be incorporated by the designer into the drawings, specification and quantities, so that they form part of the strategic decision-making process.

Traditional cost models

If the above definitions are understood it becomes clear that quantity surveyors have been using a form of modelling technique for a number of years. In their measurement for bills of quantities (BQs) they have been representing the building in a form suitable for the contractor's estimator; and when prices are applied to the measured quantities the BQ becomes a representation (or model) of the cost of the building. By altering the quantity of the measured items or changing the price according to variations in specification it would be possible to evaluate the effect on cost of manipulating certain design variables.

However, the BQ has to be prepared at a very late stage of the design process, and any information obtained from changing the quantities or price rates would come too late to avoid abortive design effort. An estimate obtained from a BQ would also be too late to give an indication of the client's likely cost commitment at the outset of the project or allow any cost control to be exercised.

Horses-for-courses

There is, therefore, an obvious need to use much simpler models at an earlier stage of design to overcome these problems.

The complexity of the model will depend very largely on the amount of information the designer can give the cost adviser, because:

- There is little point in using a complex model that takes into account shape and layout of the building if all that has been determined is an idea of the approximate area of accommodation.
- Conversely, it is wrong to use an oversimplified costing technique when the building form and specification is known and sufficient time is available to do a more thorough job.

As so often in this world the choice of technique is a case of picking the right horse for the course.

The pyramid

Figure 11.1 shows some of the more traditional models that have been developed over the years to suit various stages of the design process. The pyramid is an attempt to show that more detail is required in the structure of the model as we descend the list.

Single price rate methods

The first two model types ('unit' and 'space') are basically 'single price rate' methods of estimating cost, and those of practical application have already been dealt with in Chapter 8.

The storey enclosure method

This single price rate system is now of historical interest only, but it was the first attempt to compensate for such factors as the height and shape of buildings.

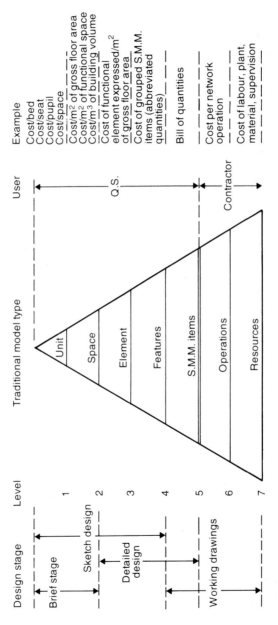

Fig. 11.1 Traditional cost modelling.

In this method the areas of the various floors, roof and containing walls were measured, each then being weighted by a different percentage and the resultant figures totalled to give the number of storey enclosure units. The rules were:

- The area of the lowest floor multiplied by 2 (or by 3 if below ground level).
- The area of the roof (measured flat).
- The area of the upper floors multiplied by 2 and plus a factor of 0.15 for the first floor, 0.30 for the second, 0.45 for the third, and so on.
- The area of the external walls, any part below floor level being multiplied by 2.

It was recommended that lifts and other engineering services should be excluded from the calculation and worked out separately. This is common practice in single price rate estimating and applies equally to the cubic and superficial methods.

The system depended heavily on the 'weightings' which were permanently built into the rules, and these were unlikely to apply equally to every building. In addition the measured units tended to be abstract, not relating to the physical form of the building nor the client's accommodation requirements. Therefore the technique suffered from deficiencies similar to those of the cube method which it was intended to replace.

The paper presenting the method (in the early 1950s) claimed that in tests it had performed far better than the other single price approaches, but lack of use meant that it was never possible to verify this claim in practice. It never achieved wide acceptance because it was quickly superseded by the elemental estimate (see later).

Comparison of single price methods

Table 5 shows each of the cube, superficial area and storey enclosure methods applied to a multi-storey office block. The actual cost of each building is known and the analysis of building A is used in each case to forecast building B. The buildings have exactly the same dimensions, except that A has seven storeys and B has eight storeys within the same envelope.

- It can be seen that the cubic method has not taken into account the extra floor and there is, therefore, a shortfall of approximately 12% on the estimate.
- The superficial method, on the other hand, has overcompensated for the extra floor by not taking into account the fact that the envelope area remains the same. It has, however, produced a more reasonable figure.
- The storey enclosure method, by considering both envelope and floor, has certainly produced an even closer estimate well within acceptable limits for this kind of early exercise.

However, in each case a change in specification, site or location would result in additional problems and call for the skill of the cost adviser's judgement to compensate for the changes. In addition, changes in plan shape and storey height

Table 5   Comparison of single price estimating models using a cost analysis of Office Block A to forecast Office Block B.

Dimensions

Office A

50 m

15 m

4.29 m

1.4 m

℄

ELEVATION

PLAN

N.B. Wall thickness = 0.25 m

Office B
Identical dimensions but eight storeys included
Therefore storey height = 3.75 m

Office A

Actual cost	= £1,758,000	Cube	= 24,000 m³	Storey height	= 4.29 m
Area	= 5,024 m²	No. of storeys	= 7	Foundation depth	= 1.4 m

Actual cost = £1,758,000   Cube   = 24,000 m³   Storey height = 4.29 m
Area   = 5,024 m²   No. of storeys = 7   Foundation depth = 1.4 m

Office B

Actual cost = £1,995,000   Cube   = 24,000 m³   Storey height = 3.75 m
Area   = 5,742 m²   No. of storeys = 8   Foundation depth = 1.4 m

---

Cubic metre

Office A analysis

$$\frac{£1,758,000}{24,000} = £73.25/m^3$$

Office B forecast
£73.25 × 24,000 = £1,758,000

Error
Underestimate of approximately 12%

Square metre

Office A analysis

$$\frac{£1,758,000}{5,024} = £349.92/m^2$$

Office B forecast
£349.92 × 5,742 = £2,009,000

Error
Well within acceptable range

Storey enclosure   **m²**

Office A analysis

Lowest floor	717.75 × 2	= 1,435.50
First floor	717.75 × 2.15	= 1,543.16
Second floor	717.75 × 2.30	= 1,650.83
Third floor	717.75 × 2.45	= 1,758.49
Fourth floor	717.75 × 2.60	= 1,866.15
Fifth floor	717.75 × 2.75	= 1,973.81
Sixth floor	717.75 × 2.90	= 2,081.48
Roof	750.00 × 1	= 750.00
External		
Walls	3,900.00 × 1	= 3,900.00
		16,959.42

$$\frac{£1,758,000}{16,960} = £103,65 \text{ SEU}$$

Office B forecast
As above   16,959.42
add
Extra floor (7th) 717.75 × 3.05   2,189.14
                    Total   19,148.56
19,149 × £103.65 = £1,984,800

Error
Well within acceptable range

would affect the cubic and superficial methods to a greater or lesser degree. These alterations, or variables, are enormously complex and difficult to assess.

Although adequate for establishing a budget, it will also be remembered that such an estimate suffers from the defect that during the development of the design it is not possible to relate the cost of the work shown on the working drawings to the estimate.

It is for these reasons that single price rate methods now tend to be rejected except for the very earliest estimates where little is known about the building form.

Elements

The model by which design cost planning has been best achieved is that of dividing the building into functional 'elements' (level 3 of our pyramid).

Elements are defined as major parts of the building which always perform the same function irrespective of their location or specification. For example:

- internal partitions always vertically divide two internal spaces;
- a roof encloses the top of a building and keeps the weather out and the heat in;
- the sub-structure transmits the building load to the sub-soil;

and so on. These elements have some relation to the design process, can be readily measured from sketch drawings, and are easily understood by all parties including the client, thus aiding communication.

Example

Suppose an external wall element costs £80,000 and the area of the wall is 1,000 m². The cost per square metre of wall (known as the elemental unit rate) is £80.00. If there are 1500 m² of the same type of wall on the proposed project then the cost will be:

$$\frac{£80,000}{1,000\,m^2} \times 1,500\,m^2 = £120,000$$

A cost index (see Chapter 14) is then used to update the prevailing price levels to the proposed tender date and the new estimate for that part of the building has been established.

Having set out a series of estimates, one for each element (known as cost targets), it is then possible to consider each element in turn instead of trying to cope with the whole building. Costs can therefore be monitored as the design develops.

Features

Cost models based on 'features' (groups of abbreviated quantities) rather than 'elements' were popular for a period among some major local authorities, but this non-standard approach was never widely used and disappeared with the decline in local authority building programmes.

Standard Method of Measurement items – the BQ as a cost model

Level 5 of the model pyramid shows the position of the BQ in terms of the detailed modelling used in traditional techniques.

There has been an attempt in the current revision of the rules for measuring BQs (the Standard Method of Measurement of Building Works) to orientate the measurement of the building to the way costs are incurred on site. However, the vast majority of items are still largely measured 'in place'; that is to say they are measured as fixed in the building with no allowance for waste and no identification of the plant and tools required to instal them. In the BQ's traditional and primary role this avoids the possible situation arising where the quantity surveyor appears to be telling the contractor 'how to do the job' or making assumptions as to his or her efficiency.

Consequently the BQ remains primarily a document for obtaining a tender in a short space of time, rather than a document for management and cost control of construction on site. It is not possible, for example, to ascertain from this document the lengths of timber required for roof joists (although this must have been known by the measurer) or the hoisting requirements for materials and components, unless the document has been annotated for the purpose.

In the case of levels 1 to 4 of the pyramid the data used are actually based upon an analysis of a BQ, which is somebody else's view (the estimator's) of what the firm needs to charge for the resources, plus profit and overheads. Because of the large number of assumptions upon which a BQ price is based it is most unlikely that the real mix of resources that will be used for the project has been determined by the estimator. The factor that allows cost techniques based on the BQ to work is the knowledge that the overall cost of the job, which has been analysed to provide data for the model, is the 'going rate' for that building.

Operations and resources

The last two levels, 6 and 7, are indeed more closely related to construction.

A contractor's 'network' outlines the activities or operations in the order that they will be undertaken on site. For example, external cavity wall brickwork, which may be one item in a BQ, will be broken down into floor levels and possibly into zones marked on the floor plan. This aids site management and the organisation of labour, plant and material.

These basic resources form the most detailed level of modelling as they are the ingredients of the production process. In fact, all the other models used for cost forecasting entail the assumption that the resource requirements of the building will just happen to work out OK (because they usually do).

Spatial costing

Some 30 years ago a variation on the superficial and elemental methods was developed under the title of spatial costing. In this method the basic concept was the cost of a room of a certain type, taking account of its floor area. There is a high correlation between the cost of room finishes, fittings, ceilings and walls, and the activity for which the room is designed.

If, at the brief stage of design, all that is available is the accommodation schedule (i.e. the list of required rooms), then it should be possible to forecast the cost of these items quite reliably as a function of the areas and, if possible, shapes of each room or space. The other major items of structural frame, foundations, external envelope and main services would then be considered separately.

The wall costs attaching to a room would be the cost of the finishings plus 50% of the structural partition, similarly with the floor and ceiling. The extra cost of external walls over and above the 50% partition costs, and the extra cost of roof over the ceiling costs, are added on a measured basis as are the other exclusions which have been mentioned.

A major problem in using this system, or any other system which depends upon BQ analysis, is the chicken-and-egg situation whereby:

- no system is going to become widely used until there is a considerable body of data to support it, but
- no-one is going to spend money on analysing BQs to provide data for a system which nobody is yet using.

The cost and difficulty of preparing detailed analyses have been contributory causes of the lack of data available for elemental cost planning. The room concept requires even more complex analysis and therefore would be even more expensive to use. The system also lacks the simplicity of the elemental form and the clear communication of what is meant and included.

Computer developments

Computer techniques now enable the economic modelling of proposed building designs in the manner already discussed to be carried out rapidly and exhaustively.

Computer models which symbolise a particular building and can build up rates from the basic resource costs are likely to provide a better solution in the future than those which simply involve analysing and manipulating old or 'historic' costs, and should be more flexible in use. However, the development of computer aided architectural design systems has added a new twist to the situation. The more sophisticated of these systems model the project in three dimensions as the architect designs it, producing not only elemental cost models but thermal performance models, day-lighting models, and so on.

It is a simple matter to write a computer program which will produce an outline elemental analysis from three-dimensional coordinates of the various corners of the building mass. The difficulty lies in associating this with cost data sophisticated and robust enough for the architect to be able to use it unassisted as the design develops, or better, to perfect a means by which the quantity surveyor or cost planner can directly interact with the architect while design work proceeds. Writing letters or telephoning will not be good enough, but facsimile and e-mail hold out more promise, while interactive computer systems can perhaps provide the best alternative to actually being in the same room as the architect.

The danger is that a model prepared from oversimplified or incorrect cost data will look just as convincing on the screen as a correct model, and will be just as easy (or even easier) to call up by somebody who does not understand the complexities involved in cost forecasting. Architects, cost planners and computer software houses must come to terms with this.

Summing-up

Although single price rate estimates have their drawbacks, these simple models of building cost are very useful for budgeting and other very early estimates where little information is available about the configuration of the proposed building. Of these the most commonly used is the superficial floor area method, and there is a good amount of data available for it. At later stages more detailed methods, such as elemental cost planning, should be used. *However, a 'cost plan' without subsequent cost control is still really nothing more than an 'approximate estimate'.*

Further reading

Fortune, C., Moores, N. & Lees, T. (1996) *The Relative Performance of New and Traditional Cost Models in Strategic Advice for Clients*. RICS Research Paper series, London.

Chapter 12
Design-based Building Cost Models

Definition

Building cost models are basically of two kinds:

- A product-based cost model is one that models the finished project.
- A process-based cost model is one that models the process of the project's construction.

It is often argued that since it is the process that actually generates the costs, the second of these two models must be the more accurate. However, the construction process cannot be modelled until the form of the building has been postulated, and therefore the process-based model has little place in the scheme of things in the early design stage.

In fact an attempt to model the construction process at too early a stage can have the effect of over-riding the design process (as set out in Chapter 10) in order to arrive at a bricks-and-mortar solution before the user criteria have been properly worked out. Construction process modelling therefore has its place at a later stage in the cost planning process, and will be dealt with in due course.

For the moment, therefore, we must be concerned with modelling the product. Such a model has to be based on data relating to finished work. As stated in Chapter 8, the simplest form of such a model takes no account of the configuration, or details of design, of the building but is based simply upon *one* of the following:

- the floor area of the proposed project (gross or net);
- the volume of the proposed project;
- some user parameter such as number of pupil places for a school or number of beds for a hospital.

It is customary to exclude site works from the single price rate calculation and to estimate them separately, since their cost obviously has no relationship to the size of the building. The engineering services are also sometimes treated in the same way (with rather less justification).

After the Second World War, and before cost planning systems became established, single price rate estimates increasingly acquired a reputation for inaccuracy, but this was because there was no follow-up process for keeping the estimate and the design in tune as the design developed. Such an estimate merely attempts to forecast that a building of a certain size can be built for a certain sum of money. It cannot analyse whether a particular design is going to meet that cost. Of course, it is possible to weight the estimate subjectively on the grounds that the proposed solution looks to be at the expensive end or the low-cost end of the market, but here we are well into the realms of guesswork.

However, although it cannot say that a particular design will achieve the required result, the single price rate approach has an important role as the first stage of a system which in the end will do that. It has the great advantage that it can be used before any design has taken place, whereas even the most simple process-based models require a tangible design as their basis.

The bill of quantities

The term 'cost model' is a relatively new one, but the traditional bill of quantities (BQ) is in fact a very good example of a product-based cost model.

Because of its insistence on measuring 'finished work in place' the BQ has often been criticised by production-oriented people (although it never purported to be a production-control document), but it does have the virtue of providing a total cost model within a single document. There are two important points in favour of the product-oriented approach to this definition of the project:

- What the client will be contracting for is finished work in place, not a production process. The model is therefore couched in contractual terms.
- The design team can define and categorise the finished work, according to an industry agreed convention ('The Standard Method of Measurement of Building Works'). This is not the case with the production processes, which depend as much on the methods of the particular constructor as on the details of the design, and which cannot easily be analysed or categorised outside the context of the particular project – there is no agreed basis for doing this.

Although the BQ is a useful cost model, it is not usually available until the design of the project is completed and it is therefore of little use for cost control of that design. Its great importance, however, is as a source of cost data for subsequent projects.

Since the BQ is a major source of cost data it is important to be quite clear about its role. It is essentially a marketing document, not a production control document, and the rates in it are prices not production costs. Although there clearly has to be a relationship between the builder's prices and the builder's costs, this really only applies at the level of the total project. The building firm cannot afford to undertake the project for less than its costs, and competition will usually ensure that it cannot charge an unreasonable profit on top of them.

This does not apply to individual pieces of work, however, because the rates in the BQ are not separately offered prices in the sense of the supermarket shelf.

You can only buy the brickwork at so much a square metre if you also buy the carpentry at whatever price is charged, so these rates are nothing more than notional breakdowns of the total price, and are made with commercial rather than cost control objectives in mind.

Very often the client's QS will object if the rate for a particular item is very different from that usually charged, but there is very little sanction to enforce an objection, except the rather impracticable one of advising the client not to accept the tender.

Elemental cost analysis

Elemental cost analysis is perhaps the best known product-based cost model, and provides the data upon which elemental cost planning is based. This technique is currently used by the quantity surveying profession at large and, in spite of a number of failings which have become apparent over the years, has made possible a degree of control over costs which was previously unknown. It is the experience gained with this system, and the attitude of mind which it has engendered, which has led the way to more fundamental approaches to cost control.

Because elemental cost analysis was developed in order to provide data for the preparation of a design cost plan it is first necessary to look at the requirements of the latter. This is done in the context of the traditional competitive tendered type of contract, which is still popular but which at the time the system was developed was almost universally used.

What are the essential features of a cost plan? The obvious requirement (common to any form of estimate prior to tendering) is that it should anticipate the tender amount as closely as possible, but there are two particular requirements that must be fulfilled before the estimate becomes a cost plan:

- It must be prepared and set out in such a way that as each drawing is produced it can be checked against the estimate without waiting until the whole design is complete. This enables any necessary adjustments to be made before the drawing is used for the preparation of quantities.
- It should be capable of comparison with other known schemes in order to see whether the amount of money allocated to each part of the building is reasonable in itself and is also a reasonable proportion of the whole. From the point of view of economical building, this second requirement is possibly even more important than the first.

It can be seen that this philosophy is especially suited to work in the public sector where, as we have seen, the only real cost criterion is comparison with what has been done before. At the time the system was developed such work dominated the building programme – it is possible that a totally different approach might have been used had profit development then been dominant.

. However, once a system has become established it tends to form the basis for future development as circumstances change, and this is what has happened with elemental cost planning, still widely used today.

How can the costs of one building project be compared with another? The first suggestion which would occur to a quantity surveyor would be to look at the priced BQs for the two contracts. Suppose we look at the summary pages of the BQs for two different schools.

	Clay Green School £	Woodley Road School £
Preliminaries	57,300	129,000
Excavation	24,240	28,980
Concrete work	129,915	272,925
Brickwork and blockwork	154,305	101,145
Masonry	10,950	—
Asphalt work	21,150	—
Roofing	74,655	39,045
Woodwork	228,675	290,895
Steelwork and metal work	189,660	197,850
Plumbing, engineering and electrical work	319,500	399,825
Floor, wall and ceiling finishes	136,725	133,275
Glazing	14,640	25,530
Painting and decoration	39,135	37,170
Drainage	43,260	33,495
Site works	109,890	79,470
	1,554,000	1,768,605
Insurances and summary items	5,250	37,950
Contingencies	30,000	30,000
	1,589,250	1,836,555

These trade totals tell us very little. We can see that the second school is dearer than the first (it is in fact larger), but it is difficult enough to try and compare the rather varied trade breakdowns without having to make repeated mental adjustments for the difference in size between the two buildings.

So a first step is to divide each trade total by the floor area of the respective schools in order to obtain comparative prices per square metre for each trade:

	Clay Green (2,000 m^2) Cost £/m^2	Woodley Road (2,500 m^2) Cost £/m^2
Preliminaries	28.65	51.60
Excavation	12.12	11.59
Concrete work	64.96	109.17
Brickwork and blockwork	77.15	40.46
Masonry	5.48	—
Asphalt work	10.57	—
Roofing	37.33	15.62
Woodwork	114.34	116.36

Steelwork and metal work	94.83	79.14
Plumbing, engineering and electrical work	159.75	159.93
Floor, wall and ceiling finishes	68.36	53.31
Glazing	7.32	10.21
Painting and decoration	19.57	14.87
Drainage	21.63	13.40
Site works	54.95	31.79
	777.00	707.44
Insurances and summary items	2.63	15.18
Contingencies	15.00	12.00
	794.63	734.62

We now have the costs on a more truly comparable basis, and it turns out that the first school is in fact the more expensive of the two. Clay Green has:

- a steel frame,
- timber pitched and tiled roofs,
- mainly brick-faced external walls with 'window holes',

while Woodley Road has:

- a reinforced concrete frame,
- felt flat roofs on composite decking,
- large wall areas of metal windows with concrete panel in-filling.

As we would expect, Woodley Road shows a far higher total for concrete work, and a lower figure for brickwork, roofing and steelwork. At Clay Green there are some asphalt covered concrete flat roofs and the crawlway floor duct is tanked in asphalt, whereas at Woodley Road the floor ducts are waterproof rendered. The glazing figures allow for the larger window area at Woodley Road, while the drainage and site works figures are consistent with Clay Green being on a larger and more rural site with a consequently greater extent of external works.

The differences in floor, wall and ceiling finishes and decoration are simply due to a higher standard of specification at Clay Green. However, while the effect of these differences in specification can be traced to some extent in the trade costs per square metre, the inclusion of parts of several elements of the building in one trade section makes it impossible to carry comparisons very far. For instance, the figures for Woodley Road for concrete work are affected by:

- the reinforced concrete frame in lieu of structural steelwork;
- some concrete walls in lieu of brickwork;
- the concrete facing panels in lieu of faced brickwork.

We do not know how much of the difference is due to any one of these causes.

Similarly, the steelwork figures are affected by the omission of the steel frame and the increased area of higher quality metal windows at Woodley Road. These are compensating differences and in fact the two figures do not reflect the scale of the variations between one school and the other in this section. Other sections are similarly affected – the rates for woodwork are almost meaningless without detailed breakdowns.

Although the figures which we have obtained are interesting, they do not enable us to make any really valid cost comparisons. We do not know whether the steel and the concrete frames are competitive in cost, and we cannot tell how much is saved by a felt roof instead of a tiled one.

In order to be able to make such comparisons we shall have to split up the BQ in a different way. We shall have to divide the work into 'elements'. An element has been defined as 'that part of the building which always performs the same functions irrespective of building type', and (we might add) irrespective of specification. What list of elements should we use? The possibilities are nearly endless. We could use as few as six, for example:

- sub-structure and ground floor, complete with finishes;
- external and internal walls, complete with finishes;
- upper floors including staircases, complete with finishes, and proportion of frame;
- roof, complete with finishes and proportion of frame;
- engineering and electrical services;
- site works.

Alternatively we could use any greater number within reason – some authorities have used over 40.

Points to bear in mind in arriving at a decision are:

- The definition of an element as stated previously. Any element chosen must be capable of being defined exactly, so as to ensure uniformity between the elemental breakdowns of any number of contracts, even if the breakdowns are done by different people.
- The element must be of cost importance.
- The element must be easily separated, both in measuring from sketch drawings and in analysing BQs.
- The list of elements chosen should be capable of bring reconciled with those used by others, for comparison purposes.

A cost planner who does not have access to the records of a very large firm or other organisation will frequently need to make comparisons with analyses published in *Building* magazine and elsewhere, or with analyses obtained from the Building Cost Information Service of the RICS. As an example of the need for standardisation, in most published forms of cost analysis the cost of parapet walls and copings is included under 'roof'. A cost planner might prefer to include it under 'external walls' (it would make the analysis of a traditional BQ easier), but would then be unable to compare the figures for these two elements with the published information, as the basis of measurement would be quite different.

The Standard Form of Cost Analysis

As the use of elemental cost planning increased, differing forms of cost analysis were developed by those independent authorities and firms who were building up their own cost records, and by journals which published cost information for the benefit of their readers. The weekly magazine *The Architects' Journal* was one of the pioneers in this field, and produced a detailed set of rules for their published analyses, while the RICS Building Cost Information Service and others used somewhat different rules.

As previously stated, it was obviously desirable that a uniform set of rules should be established, so that users could benefit fully from cost data prepared outside their own organisations. The RICS therefore set up a working party to standardise cost analyses. This proved to be difficult because, in practice, it is not possible to define a set of totally independent functional elements which can be related to a BQ. Any standard cost analysis therefore becomes a compromise between independent functions on the one hand and ease of producing the data from traditional documentation on the other.

In December 1969 the first Standard Form of Cost Analysis (SFCA) was published by the Building Cost Information Service of the RICS. In addition to its sponsorship by the RICS, the SFCA was also supported by the chief quantity surveyors of all the main government departments which were concerned with building (and this was at a time when these bodies directly controlled a major part of the non-housing building programme).

It was this wide measure of support which gave the SFCA such importance, as anybody using a different format would soon have become isolated from the cost experience of the rest of the quantity surveying profession. Thirty years later the SFCA is little changed, since it works reasonably well and any alteration would have the effect of making the comparison of old and new projects unnecessarily difficult.

As an example of a practical elemental breakdown, the headings in the current SFCA are as follows:

1. Sub-structure
2. Super-structure
    2.A.   Frame
    2.B.   Upper floors
    2.C.   Roof
             2.C.1   Roof structure
             2.C.2   Roof finishes
             2.C.3   Roof drainage
             2.C.4   Roof lights
    2.D.   Stairs
             2.D.1   Stair structure
             2.D.2   etc.
    2.E.   External walls
    2.F.   Windows and external doors
             2.F.1   Windows

          2.F.2   External doors
      2.G.   Internal walls and partitions
      2.H.   Internal doors
3.   Internal finishes
      3.A.   Wall finishes
      3.B.   Floor finishes
      3.C.   Ceiling finishes
          3.C.1   Finishes to ceilings
          3.C.2   Suspended ceilings
4.   Fittings and furnishings
      4.A.   Fittings and furnishings
          4.A.1.   etc.
5.   Services
      5.A.   etc.
6.   External works
      6.A.   etc.

This summary is sufficient to enable the principle to be understood, and full details of all the sub-divisions are given in Appendix A. Most other forms of elemental analysis are basically similar, but in order to be successful they must incorporate the same sort of hierarchical principle.

It will be seen that the SFCA can be used at different levels of generality:

- the six element groupings Sub-structure, Super-structure, Internal finishes, etc.;
- the elements themselves 2.A., 2.B., 2.C., etc.
- the sub-elements 2.C.1., 2.C.2., 2.C.3., etc.;

but because the detail is grouped in this hierarchical way an analysis at level 1 into six items only will be quite compatible with a fully detailed analysis at level 3 of another project.

The construction of the list therefore required a careful selection of elements, each of which was significant on its own but which could form part of a larger significant group. Note, however, that the third level of detail is no longer used by the Building Cost Information Service in its published analyses – a tacit admission of the unreliability of sub-elemental analysis referred to later in this chapter.

One or two forms of analysis which have been used differ by:

- including finishes, windows, and external doors in the external walling element;
- including finishes, internal doors, and partitions in the internal walling element.

This is more logical but means splitting finishes between external and internal walls, which is difficult to do when analysing a traditional BQ.

Using elemental analyses

We can see that if the costs per square metre of Clay Green and Woodley Road schools could be expressed in terms of these elements it would enable us to find the answers to the questions which we were asking:

- Which is the cheaper frame?
- How do the two roofs compare in cost?

These answers would then enable us to cost plan a third school basically similar to Woodley Road but with a steel frame.

The Standard Form, CI/SfB and Uniclass

It was a source of disappointment to many people that in the BCIS Form the elements themselves, their grouping and their coding were not the same as in Table 1 of the CI/SfB system used by the architectural profession for coding and classifying design information. It would have been very convenient for the architect to have a record of typical costs filed with his design information.

However, the incompatibility of the systems was not quite such a disadvantage as might appear. We have seen that the so-called 'elemental costs' which we obtain at present are not true costs at all, but are only a breakdown of a BQ in which money may have been allocated to the various parts of the work in a fairly capricious manner, quite apart from the overall level of pricing of the BQ itself. It would therefore be dangerous to detach these 'costs' from the analysis and index them for the use of an architect as though they were scientific data like thermal insulation values. One day it may well be possible to do this, but at present it is vital that this cost data should only be used by a qualified cost planner who has the experience and knowledge necessary to assess and manipulate it.

The united classification for the construction industry *Uniclass*, published in 1998 to replace CI/SfB, contains an element table much closer in form to the SFCA and which is a compromise between the needs of librarians and those of the cost planner. It has a flexible form which allows the elements to be ordered in different ways for different purposes. There is a one-to-one mapping between the SFCA elements and the Uniclass elements which will allow elements to be coded in both systems.

Preambles to the SFCA

In the SFCA there is rather more information required to be given about the size and nature of the building than was previously customary. The gross floor area is measured in the normal way, that is the overall area at each floor level within the containing walls, but the basement floors (grouped together), the ground floor and the upper floors (grouped together) are each required to be shown separately.

The definition of 'enclosed spaces' means that open entrance areas, etc., are excluded from the gross floor areas, although even a light enclosing member such as a balustrade will suffice to include the area. It is possible that doubt might arise when a wall becomes so pierced by blank openings that it no longer acts as an 'enclosing wall'; obviously a mere series of columns does not meet the definition.

Although 'lift, plant, tank rooms, and the like above main roof slab' are to be included in the gross floor area, we have the option of excluding these if we are prepared to allocate their costs to 'builder's work in connection'. If the plant room is only required because of the existence of a particular service it seems quite logical to adopt the second course.

Roof and wall areas in the preambles to the SFCA

These are both measured gross over all openings, etc., the roof being measured on plan area. As it is 'walls of enclosed spaces' which are required, parapets, gable ends of unused roof spaces, etc., would not be included in the wall area. The roof, on the other hand, is measured across overhang and would presumably include roofs over open entrance porches and other areas which do not count as 'enclosed spaces'. If there are substantial areas of open covered way it would probably be better to exclude them from the elemental analysis altogether and deal with them under 'site works'.

The wall area is not shown on the Form, except in calculating the wall-to-floor elemental ratio, which is the wall area divided by the floor area. The lower the ratio the more economical the design. For this purpose the wall area is normally measured across openings.

Interdependence of elements

Although it is quite easy to define a cost planning element it is more difficult actually to divide up a BQ into elements which comply with this strict definition. A major difficulty is that an indivisible building element may have several functions (some of which it shares with other elements). For example, 'external walls' may have all or some of the following functions:

- keeping out the weather;
- thermal insulation;
- sound insulation;
- supporting themselves (dead loads, wind loads);
- supporting floors and roofs;
- transmitting light and ventilation (curtain walls).

If an external wall only performs a few of these functions then it is obviously unreasonable to compare its cost with that of a wall which performs a greater number. It is therefore usually necessary to refer to the 'frame' and 'window' elements in order to arrive at a true indication of the wall's cost performance, and similar cross-references may be required when costing other elements.

Preliminaries and insurances

These may be either shown as separate costs per square metre (treating them, in fact, almost as extra elements) or allocated proportionately among other elements. This is an important point and deserves some consideration.

First, preliminaries and insurances. The cost of any or all of the following items (mostly at the contractor's discretion) may be included in the preliminaries or insurances sections of a priced BQ for a new project:

- huts, temporary buildings, latrines;
- canteen, mess-rooms, site catering staff, welfare;
- huts for clerk of works, site architect, consulting engineers, and quantity surveyor, and attendance on these people;
- huts for contractor's own supervisory staff and (on a large project) sub-contractors' supervisory staff.
- mechanical plant, including tower cranes, excavating and concreting plant, lifts, dumpers, etc.;
- scaffolding;
- non-mechanical plant and small tools;
- water for works;
- temporary electricity supply;
- consumable stores;
- temporary fencing and hoardings;
- temporary roads and standings, car parking spaces;
- health and safety requirements;
- cost of agent, foreman and other site supervisory staff;
- cost of timekeeper and site clerks;
- cost of security staff;
- security lighting;
- heating of building for winter working and for drying out;
- temporary weather protection;
- attendance on sub-contractors and artists;
- National Insurance payments;
- superannuation;
- guaranteed week (wet time), travelling time and expenses, subsistence and lodging;
- redundancy payments;
- training levies;
- anticipated increases in costs of labour or material;
- bonus or other supplementary payments;
- making good damage and defective work;
- fire insurances, third party insurance and any other insurances required by the client;
- public liability or contractor's all-risk insurance;
- head office expenses (overheads);
- profit.

However, almost any of these items (and certainly any of the major ones) may be included in the rates for the building work instead of being shown separately.

Some contractors do not price preliminaries at all while others do price preliminaries, but in such a way that it is impossible for the quantity surveyor to find out what is or is not supposed to be included. Sometimes the contractor may have second thoughts about a tender at the last minute, and may adjust it by adding a lump sum to preliminaries, or taking one off.

Thus any large differences between the amounts of money inserted against the preliminaries items on one project and on another are less likely to be caused by genuine contractual differences (site conditions, access, etc.) than by the different pricing habits of the two contractors – remember that the allocation of costs within a contract BQ is done largely for commercial purposes.

Some of the principal items such as profit, supervision, scaffolding, plant and overheads could affect the level of pricing of the work sections by 15 to 20%, according to whether or not they are included in preliminaries.

As these pricing habits are to some extent regional it may be possible for the quantity surveyor's department of a local authority, or for a firm of cost planners whose work is confined geographically, to consider preliminaries and insurances as a separate element with some degree of consistency. However, it would be safer on the whole to add preliminaries and insurances to each element as a percentage in order to give a common basis of comparison.

If we refer back to the summaries of the two schools, Clay Green and Woodley Road, we can see that the work section prices for the latter appear low by comparison because preliminaries and insurances have been priced more fully.

This advice, of course, relates only to the analysis of BQs and to the early stage estimates for a new project based upon such data. When preparing a more detailed estimate for a major new project preliminaries and insurances have to be considered on their merits. If the new project has some abnormal feature such as a difficult site or an uneconomically short contract period it will be necessary to give special consideration to these matters from the start.

Contingencies

We have also to consider contingencies. Unlike preliminaries, the contingency sum is an arbitrary amount decided by the client or the design team. It is not really part of the contractor's tender but is an amount the contractor is instructed to add to his tender in order that there may be a cushion to absorb unforeseen extras. It normally has no effect on the level of pricing of the BQ, and is better treated as a separate element rather than as a percentage on the remainder of the work.

Analysis of final accounts

Upon first consideration it might seem a good idea to analyse the final account instead of the tender, since this will give a more accurate picture of the actual cost of the building. The objections are twofold:

- It would be much more difficult to analyse both BQ and variation account than to analyse the BQ alone.
- The analysis would not be available until perhaps 3 or 4 years after a tender analysis and so would only be of historic interest.

However, these difficulties do not seem to be insurmountable, and the convention that it is the tender that is analysed probably owes a lot to the fact that the practice of cost analysis started in the days of public sector building, when attention was focused on forecasting tenders rather than final accounts. Although the differences between the tender and final account are not usually great enough to invalidate an analysis obtained from the BQ, this is not always the case.

Cost analysis of management contracts

Management contracts and the like pose a difficult problem. There is little point in analysing the master estimate, since there is no contractual commitment to this – it is itself part of a cost planning system and may well have been prepared on an elemental basis. On the other hand, by the time all the costs are known we would be effectively analysing a final account, with all the problems just mentioned. However, management contracts imply a production orientation to the project, and it is perhaps the resource-based cost information obtainable from these projects which is more useful than the product-based costs.

System

Whatever methods of cost breakdown and whatever methods of cost planning are chosen they must be adhered to rigidly. Otherwise, not only are the figures useless for reference purposes but the whole idea of working to a system is lost.

The whole technique of cost analysis depends upon working in accordance with a fixed method; this is why standard forms should be used. There is plenty of scope for rough working, but the forms must be used at all vital points, so that there is no possibility of preliminaries (for instance) being left out because the cost planner who did the previous estimate believes in showing them separately, whereas the person using this estimate for reference supposes that they are included in the rates.

In most cases the forms and instructions issued by the BCIS can be used, thus ensuring compatibility with information obtained from other sources.

Preparation of an elemental cost analysis

Cost analyses are most usually carried out today using a computer. Special software can be bought, as explained in Chapter 2, but many cost planners find that commonly available spreadsheet packages enable them to construct their own cost analysis system very easily.

The aim of cost analysis is to provide data for use in elemental cost planning; as little time as possible should be spent on it consistent with obtaining a fair degree of accuracy. Meticulous allocation of trivial sums of money, or the identification of insignificant changes in specification, should be avoided. If there is no time to prepare a full analysis an outline analysis will be better than nothing and may take less than an hour to prepare.

For the preparation of an elemental cost analysis we shall require:

- a priced BQ;
- a drawing showing plans and elevations;
- a list of elements.

Each item in the BQ has to be allocated to one or more elements until every item has been dealt with and the elemental totals will equal the total of the tender.

Once an office has adopted a certain form of analysis it will be possible to prepare the BQs with subsequent analysis in mind; this will ease the task of the analyser considerably and will make it unnecessary to refer back to the taking off.

It is not necessary to depart radically from the usual order of billing as long as the main elements can be kept separate within each trade or section of trade. For instance, the 'sawn softwood' section of woodwork could be billed under headings of:

- roof timbers;
- upper floors;
- stud partitions;

and the 'reinforced concrete' section of the concrete work could be separated into:

- sub-structure;
- frame;
- upper floors;
- roof;
- staircases.

This should not involve lengthening the BQ greatly as there will be very little duplication of items between different elements; the concrete work is the only section where this should occur to a significant extent.

A BQ prepared in this manner will be useful not only for analysis, but it will be convenient for interim valuations and for the contractor's use in site organisation generally and in calculating performance-related payments to operatives.

The so-called Northern system of taking off, where the BQ is written straight from the dimensions, lends itself to the preparation of a sub-divided BQ of this sort.

Any further breakdown of the tendering BQ into a completely elemental format is very unpopular with builders' estimators, since design cost planning elements are of little significance to them and a good deal of re-arrangement into trade order has to be carried out by them in sending out to sub-contractors and suppliers.

The Hertfordshire Bill of Quantities

This is a convenient point to mention the ingenious BQ which was pioneered by Hertfordshire County Council before computers were available, in which each trade of each element was printed on a separate sheet and coded. The BQ could thus be shuffled into either elemental or trade order at will. The usual arrangement was for the Bill to be sent out in trade form for tendering, and put into elemental form for subsequent use.

The Master Bill

Another development was the Master Bill for tendering, for use when a contract consisted of several buildings, each of which required a separate Bill which in turn was divided into trades and/or elements. In order to avoid inundating the contractors who were tendering with masses of paper the section Bills were re-abstracted into a Master Bill which was used only for tendering. This Master Bill was in strict trade order and contained no sub-headings or sections apart from those required by the Standard Method of Measurement.

The additional expense of preparing such a Bill may have been worthwhile because busy contractors naturally prefer a 200 page Bill to a pile of Bills containing up to a thousand pages, and keener tendering will probably result.

By using this method it is possible to have the section Bills divided into elements without bothering about trade divisions, which are only really required for tendering purposes.

Computer Bills of Quantities

Both of the above arrangements become very easy to do with the advent of BQs prepared using specialist computer systems, and further refinements become possible. Providing that all the dimensions are given elemental code references, the BQ can be sent out to tender in traditional form and after tenders have been received the computer will be able to print an elemental version of the Bill.

In some systems the BQ rates are also fed into the computer so that the elemental breakdown will be completely priced as well.

Definition of terms

Before going any further there are a number of terms which must be defined in order that the processes of cost analysis and cost planning may be understood.

- *Elemental cost* is the cost of the element expressed in terms of the superficial area of the building.
- *Elemental unit quantity* (sometimes also called quantity factor) is the actual quantity of the element, expressed in square metres for such elements as floors, roof, walls, finishes or in terms of number of elements where this is not practicable.

- *Unit cost* is the cost of the element expressed in terms of the element unit quantity, e.g. 1,000 m² of internal walling costing £20,000 gives a unit cost of £20.00/m².
- *Elemental ratio* is the proportion which the unit quantity of one element bears to that of another. A commonly used example of this is the ratio of external wall area to gross floor area.

When analysing a BQ there are three different ways in which the analyser may be helped to deal with the items:

- The description of the item may indicate the element to which it belongs, in which case there is no necessity to refer to the original taking off.
- The item may be too trivial to spend time on, in which case it may be allocated as seems most obvious or as is most convenient.
- It may be necessary to refer to the taking off in order to allocate the item correctly. Since it would take far too long to do this for every item in the bill, analysers must use a good deal of discretion about this.

In cases of doubt they would be influenced very much by the cost importance of the item.

Element unit quantities

Unfortunately a simple analysis of building cost into elements will not satisfy all our requirements. It will give us a money total for each element and we can divide each total by the floor area to get the elemental cost per square metre, but it does not give us the unit quantities or the unit costs. We need these if the analysis is to be of much use. Some commonly used element unit quantity factors are set out below as a guide; where 'none' is marked the elemental cost per square metre is the only basis of cost comparison.

Work below lowest floor finish	Area of lowest floor
Frame	Area of floors relating to frame
Upper floors	Area of upper floors
Roof	Area on plan of roof measured to external edge of eaves, but excluding area of rooflights
Rooflights	Area of structural opening measured parallel to roof surface
Staircases	Number and total vertical rise of staircases or Area on plan
External walls	Area of external walls excluding window and door openings (Basement walls to be given separately)

Windows	Area of clear opening in walls
External doors	Area of clear opening in walls
Internal walls and partitions	Area of internal walls excluding openings
Internal doors	As for external doors
Ironmongery	None
Wall finishes	Area of finishes
Floor finishes	Area of finishes
Ceiling finishes	Area of finishes
Decorations	None
Fittings	Often none, but where appropriate the total length of benches, number of tables or other details might be given
Plumbing and hot water services	Number and type of sanitary fittings, number of hot and cold draw-offs
Heating services	Heat load in kW, cubic capacity of accommodation served
Gas services	Number of outlets
Electrical services	Number of points, total electrical load.
Special services	Such information as will indicate the extent of each service (e.g. for lifts the number, capacity and speed of each and number of stops should be given).
Drainage	None
External works	None

As well as giving the consolidated unit quantities and costs for each element, it is common practice to sub-divide them into categories of different construction or finish and give quantities and cost for each.

Example

For example, for a building of 2,400 m^2 of total floor area:

ELEMENT: Upper floors
    Total cost of element: £85,525
    Cost/m^2 of total floor area (elemental cost) £35.63
    Unit quantity and cost
    Element unit quantity   1,000 m^2
    Unit cost/m^2 £85.00

Sub-division		£
550 m² of 150 mm rc slab	at £82.50 =	45,375
350 m² of 225 mm rc slab	at £105 =	36,750
100 m² of 25 mm softwood boarding on		
175 × 50 mm joists	at £34 =	3,400
		85,525

While these sub-divisions are very useful as explanations of how the total cost is affected by specification, it must never be forgotten when using them that the greater the detail in which a priced BQ is analysed the less reliable are the results. It is more than likely that a BQ for the same job priced by another builder would give completely different figures at this level of breakdown, although quite similar in total.

The overall unit costs will be found particularly useful when preparing early estimates of cost before specification details are available.

Obtaining unit costs from elemental analyses

In order to obtain unit costs as well as elemental costs the analysis has to be more elaborate than if elemental costs alone are being recorded. This is especially so if we want to keep separate costs for the different forms of structure or finish within each element. There are two problems to solve:

● the separation of the costs within the element;
● the recording of areas and other quantities.

Such things as areas of walls, floors and finishes can be obtained most easily from the BQ. However, sizes of window and door openings may be difficult to get from this source (particularly where there are fanlights, sidelights or windows glazed directly into frames) and it may be necessary to refer to the taking-off dimensions or the drawings.

There is also the difficulty of areas which occur more than once, for instance:

● The areas of concrete in floors will be duplicated by the formwork areas and these must not be added in again.
● However, the separate areas of concrete floors and hollow pot floors will require to be added together.

Thus either the analysis needs to be done by somebody technically qualified or else very clear procedures have to be laid down.

No attempt need be made to separate constructions which differ only in detail (such as 100 mm, 125 mm and 150 mm floor slabs), because the aim is to obtain overall unit rates for basic constructions, not 'Bill rates' for individual items.

The analysis in its final form

This is one of the places where a standard form, preferably the SFCA form, should always be used. It is not necessary to fill in too much specification detail if the analysis is for office use, as it should be possible to refer to the contract papers if anything more than a very broad outline is required.

There should be a reference number for the analysis so that a list of analyses can be kept, and there should also be a cost index value so that allowance can be made for changes in market prices when comparing with past or future jobs.

Note that elemental costs continue as a running total but unit costs cannot be carried forward, as a total of them would be meaningless.

The BCIS Online system

As well as providing printed elemental cost analyses, the BCIS offers a computer-based service whereby their database of analyses can be examined from a distance, and any interesting examples downloaded to the cost planner's own computer for use in the cost planning of new projects.

This service also provides the facility to amend the BCIS analyses for the cost implications of a change in tender date or UK region of construction, so enabling comparison between jobs carried out at different times or in different places. The system has been described in detail in Chapter 2.

Design cost parameters

According to the Concise Oxford Dictionary a parameter is 'a quantity which is constant in a particular case considered, but which varies in different cases'.

It is now necessary to look at the factors which influence the components of the building in terms of area, number and size, as well as their quantity in terms of cost. Unfortunately insufficient research has been undertaken to date to give clear indications of the degree to which changes in the parameters of the building (or by implication its model) will affect the cost of that building. There is, however, a very great depth of knowledge gained by practitioners which provide us with some general 'rules of thumb'.

In some cases we can be quite specific about how cost varies. For example, if we change the shape of a single-storey building so that the area of the external brick cavity wall is increased, then we can be sure that, all other things being equal, the wall cost will probably have increased in direct proportion to the increased area. Similarly, if the quality of facing bricks is increased and the shape of the building remains fixed, then the wall cost will have increased by the extra material cost of providing the better specification.

Whilst we can probably rely on this type of simple wisdom for small brick buildings it may not be adequate for dealing with more complex multi-storey framed and curtain-walled structures. If we change the shape or height of the building, it may not be just the extra quantity and quality that we have to pay for, but also indirect costs such as:

- different lifting equipment;
- improved fixings to deal with increased exposure;
- access, and manoeuvrability and dispersal of plant on site.

A particular difficulty lies in producing rules of general application, rather than in relation to one constrained set of circumstances, and indeed there is no real agreement that such rules exist. Very often the answer seems to depend upon the methods that a given builder normally uses.

Using our existing knowledge we can, however, establish some starting principles which could be the foundation for any further cost research, but which meanwhile can be drawn upon in developing a design. We can view the parameters at two levels:

- The form of the building itself (or as it is sometimes called its 'morphology'), where we can study the effect of shape and height on cost.
- The major components of the building and the factors which influence their size, quantity and cost.

Building shape

The building shape has its major impact on:

- the areas and sizes of the vertical components such as walls, windows, partitions, etc.;
- perimeter detailing such as ground beams, fascias, and the eaves of roofs.

It would seem obvious that the building that has the smallest perimeter for a given amount of accommodation will be the cheapest as far as these items are concerned.

However, the shape that has the smallest perimeter in relation to area is the circle and this does not very often produce the cheapest solution for the following reasons:

- The building is difficult for the constructor to set out.
- Curved surfaces, particularly those incorporating timber or metalwork (e.g. in joinery or formwork) are expensive to achieve.
- Circular buildings seldom produce an efficient use of internal space, as inconvenient odd corners are generated between partitions and external walls.
- There is a tendency for circular buildings to generate non-right-angled internal arrangements.
- Standard joinery and fittings are based upon right angles and will not fit against curved surfaces, nor into acute-angled corners.
- The circle does not normally allow efficient use of site space.

In these circumstances it would appear that the right-angled building which has the lowest perimeter will provide the best answer. This shape is of course the square.

Now although a compact square form is generally recognised as the most economical solution because of its reduction in cost of external vertical elements (and the lowest area of external wall for heat loss calculations), there are some important qualifications to be made:

- Where the square plan produces a very deep building requiring artificial ventilation, air-conditioning and lighting which would not otherwise have been needed, then the shape may become uneconomic.

  It is only the single-storey building with its opportunities for natural top-lighting and ventilation which may be regarded as a partial exception to this rule, although really satisfactory arrangements for top-lighting and venting are likely to be expensive in first cost and (particularly) maintenance.

- Where there is a high density of rooms the use of the external wall as a boundary to the room helps to reduce the amount of internal partitioning required.

  It is therefore sometimes preferable to elongate the building so that rooms can be served from either side of a spinal corridor, rather than have a deep building resulting in a complex network of corridors to serve all rooms plus the possibility of artificial ventilation to those that are internal.

  A real-life case concerns a high-tech building where the internal offices with no natural lighting were so claustrophobic that no-one would work in them and they were used as stores for rubbish; a small saving in cost/m^2 of floor area had produced a lot of floor area that was in fact unusable, and a poor bargain.

- A given amount of accommodation housed in a square multi-storey block may be much more expensive than the same accommodation housed in a less compact two-storey block, for reasons to be discussed later.

- On a sloping site involving cut-and-fill it may be more sensible to provide a long building running with the contours rather than a square building which would cut more extensively into the site.

These are just a few of the qualifications which need to be made when talking about the efficiency of shape.

Like a number of rules-of-thumb, this one was developed for traditional buildings in the UK, and two exceptions show how dangerous it is to regard such rules as universal laws:

- Modern high-tech buildings normally require large floor areas, air-conditioning and artificial lighting whatever their shape.
- Part of the reason why external walls are expensive in countries like Britain is because they need to have a good thermal performance. In warm countries this is not the case, and the ability to obtain natural ventilation cheaply may lead the cost planner to try to *maximise* the perimeter of the building.

In any case, the shape of the building is often dictated by the site boundaries, topography and orientation, and the degree of choice is therefore rather limited. If the national construction programme tends towards redevelopment rather than the exploitation of green field sites, then ideals of building shape can become fairly meaningless. There have, however, been a number of attempts to measure the cost efficiency of a building shape, and some simple examples are listed below:

- Wall/floor ratio
  This is perhaps the most familiar of all the efficiency ratios but it can only be used to compare buildings with a similar floor area and does not have an optimum reference point such as those below.

- $\dfrac{P - Ps}{Ps} \times 100\%$

  (J. Cooke)
  Where P = perimeter of building
  $\quad$ Ps = perimeter of square of the same area
  This simple formula relates any shape to a square which would contain the same area thus providing a reference point for shape efficiency.

- Plan compactness or POP ratio (Strathclyde University)

  $\dfrac{2(\pi A)^{\frac{1}{2}}}{P} \times 100\%$

  Where P = perimeter of building
  $\quad$ A = area of building
  In this case the point of reference is the circle (a square would have a POP ratio of 88.6% efficiency and yet it is probably the best cost solution in initial cost terms).

- Mass compactness or VOLM ratio (Strathclyde University)

  $2\pi \dfrac{[(3V/2\pi)^{\frac{1}{3}}]^2}{S} \times 100\%$

  Where V = volume of hemisphere equal to volume of building
  $\quad$ S = measured surface area of the building (ground area not included)
  This formula chooses a hemisphere as the point of reference for considering the compactness of the building in three dimensions.

- Length/breadth index (D. Banks)

  $\dfrac{p + \sqrt{(p^2 - 16a)}}{p - \sqrt{(p^2 - 16a)}}$

  Where p = perimeter of building
  $\quad$ a = area of building
  In this index any right-angled plan shape of building is reduced to a rectangle having the same area and perimeter as the building. Curved angles can be dealt with by a weighting system. The advantage here is that the rectangular shape allows a quick mental check for efficiency. As these formulae are only for guidance purposes this index is probably sufficient for early design advice.

- Plan/Shape Index (D. Banks)

  $\dfrac{g + \sqrt{(g^2 - 16r)}}{g - \sqrt{(g^2 - 16r)}}$

Where g = sum of perimeters of each floor divided by the number of floors
r = gross floor area divided by the number of floors

This is a development of the previous index to allow for multi-storey construction. In effect, the area and perimeters are averaged out to give a guide as to the overall plan shape efficiency.

While the above indices are useful they obviously have severe limitations as they consider only those elements that comprise the perimeter of the building, or in the case of VOLM the perimeter and roof.

However, the repercussions of shape on many other major elements are considerable. For example, wide spans generated by a different plan shape may result in deeper beams, which in turn demand a greater storey height to give the same headroom, and thus will affect all the vertical elements. These implications need to be represented in any advanced model of building form, and an awareness of knock-on cost effects of this kind must be part of the cost planner's knowledge.

Size of building is another important factor in cost efficiency. The larger the plan area for a given shape, the lower will be the wall/floor ratio. To take the example of a single-storey square building with a height of 5 m:

size of square	floor area	wall area	w/f ratio
10 m × 10 m	100 m²	200 m²	2.00
20 m × 20 m	400 m²	400 m²	1.00
30 m × 30 m	900 m²	600 m²	0.67

Although the previous points which have been made, concerning artificial lighting and ventilation in particular, may counteract this saving in envelope area, other savings that result from larger plan area buildings make the grouping of functions into a single large building cost-attractive.

As already pointed out, external walls and their finishes in non-tropical countries need to have a high performance, as they are exposed to the weather and tend to be expensive, so that the savings to be obtained from a reduction in external wall area are considerable.

Conventional terraced housing is a good example of cost savings due to grouping; by taking three detached houses and making them into a terrace you save:

- two brick-faced flank walls;
- two lengths of foundation for these walls;
- four roof verges, downpipes, and so on;
- two other brick-faced flank walls become cheaper party walls.

Further savings could be made by having deep narrow-fronted terraced houses rather than wide-fronted ones.

Another advantage of large buildings containing grouped accommodation is the sharing of common facilities. The more accommodation units that can be grouped around the central service and circulation core of a building, the less will be the cost of the service and access facility for each unit. This applies both to residential buildings and to offices and the like, although over-centralisation can involve long circulation routes and corridors, and problems with fire escapes, and in the case of offices can make the sub-division of spaces for letting difficult. However, given the choice most of us would prefer to live in a detached house rather than a terraced one, or to work in an office which did not involve a long walk to an over-used lift.

Here we come face-to-face with a nasty fact: once any obvious extravagances in a scheme have been cut out, further savings nearly always involve a reduction in the quality of the building environment. A conscious choice has to be made; sometimes the saving is clearly worthwhile, while in other cases the net benefit is much more questionable. The experience and knowledge of all members of the design team must be called upon here, but most importantly the client should be closely involved in the process of choice, and the resultant trade-offs.

Height

Here it is possible, and indeed desirable, to be dogmatic. Tall buildings minimise land costs in relation to floor area, but are invariably more expensive to build than low-rise buildings offering the same accommodation, and the taller the building the greater the comparative cost. The only partial exception to this rule is that the addition of a further storey or storeys to a tall building in order to make the best use of lifts or other expensive services may slightly decrease the cost per storey, but this does not invalidate the general rule.

Cost problems of tall buildings

What are the reasons for the high cost of high buildings?

- The cost of the special arrangements to service the building, particularly the upper floors. Apart from the necessity of providing sufficient high-speed lifts it is necessary to pump water up and to break the fall of sewage and other rubbish coming down. Complete service floors often have to be provided at intervals of 10 or 15 floors to deal with these problems.

- Special ventilation and lighting arrangements are needed because of the impossibility of providing adequate light wells in a tall building.

- A high standard of fire-resistant construction and practicable escape arrangements are required.

- The necessity for the lower part of the building to be able to carry the weight of the upper storeys, which obviously makes it more expensive than if it were carrying its own weight alone.

- The structure of the building and its cladding will have to be designed to resist a heavy wind loading, a factor which hardly affects a low building at all. Experience with many of the tall buildings of the last 30 years has shown how demanding is the required standard of windows, wall panels, etc., at high levels, and how expensive is the failure to meet these standards. One is talking about a very different price range indeed from similar components for low-rise construction.

- The cost of working at a great height when erecting the building. There is the cost of:
  - Hoisting all materials and operatives to the required level.
  - The time spent by operatives going up to their work and down again at the beginning and end of each day and at break times.
  - The extra payments for working at high altitudes and all the safety requirements which this entails.
  - The bad climatic conditions for working at many times of the year.

- The increased area occupied by the service core and circulation. As the height of a building increases so it needs more lifts, larger ducts, wider staircases, etc., and these installations take up more and more of the lower floors, so cutting down on the usable area. It is possible to imagine a building so tall that the whole of the ground floor is occupied by vertical services; adding extra floors to such a building would produce no increase in usable area at all!

- The cost of dealing with the effects on neighbouring properties, such as rights of light, and the considerable costs of overcoming planning objections.

Many of the above factors will also influence the running and maintenance costs – such items as window cleaning, repainting and repairs to the face of the building will all be much more costly than similar work to a low-rise structure. Therefore high building should never be considered favourably on cost grounds unless the saving in land costs due to the smaller site area in relation to accommodation will pay for the considerable extra building costs. Land values must be high for this to occur, since a tall building needs a lot of space around it and the reduction in land requirement is not proportional to the reduction in plan area.

An increasing problem today is the near impossibility of dealing with the car-parking needs of a tall building within the plan area of the site, except at an excessive cost.

Costs of single-storey buildings

Just as high building is not usually economical neither is single-storey building, but the exceptions here are much more numerous and important, for instance:

- Where large floor areas free from obstruction by wall or columns are required it is more economical to build horizontally rather than vertically, since to provide loadbearing floors over such areas would be far more difficult and expensive than a roof.
- Similarly, where very heavy floor loadings are required it is cheaper to build floors resting on the ground than high-performance suspended floors over other storeys.

- Single-storey temporary or sub-standard buildings where the low-cost foundations and structure cannot be made capable of supporting a further storey can nevertheless be an economical solution.

However, both for retailing and for manufacture there are so many user advantages in single-storey accommodation that this type of building is becoming very common. One of the problems in some areas is in finding a reasonably level site for a large building of this kind.

Cost advantage of low-rise buildings

The reason for the relative economy of two- or three-storey buildings compared to single-storey is that:

- One roof and one set of foundations will be serving two or three times the floor area.
- The walls or frame will be capable of carrying the extra load with little or no alteration.
- In domestic construction it will be possible to use cheap timber-framed upper floors which will help the comparison still further.

Once the building exceeds this number of storeys various factors make it difficult to attain the low costs possible with two-storey construction:

- It is less often possible to dispense with a separate frame.
- The frame itself must be more substantial.
- Lifts, fire-resisting construction and other expensive measures are required.

Optimum envelope area

The 'envelope' of a building, that is the walls and roof which enclose it, forms the barrier between the inside and the outside environments. The greater the difference in these environments the more expensive this envelope will be, but it is always at least a significant factor in the cost of constructing and running the building.

We have already seen that a square building is inherently economical in wall area, but the total envelope/floor area ratio will also depend upon the number of storeys that are chosen for the accommodation. For example, imagine a large single-storey building. If we arranged the same accommodation on two floors we should reduce the roof area by more than the consequent increase in wall area, so that the total envelope area would be reduced.

The same thing might happen if we arranged the same accommodation on three floors, but if we continue making the building higher and higher, whilst retaining the same gross floor area, the process eventually reverses and the increase in wall area becomes greater than the roof area saving. This reversal happens quite slowly, so that the envelope area changes very little over a range of several possible storey arrangements close to the optimum. It is obviously useful to know what this optimum is as a design guideline. We can use a formula to calculate this optimum for a square building thus:

$$N\sqrt{N} = \frac{x\sqrt{f}}{2s}$$

N = optimum number of storeys
x = roof unit cost divided by wall unit cost
f = total floor area (m^2)
s = storey height (m)

If the desired width in metres (w) is known the formula for a rectangular building is:

$$N^2 = \frac{xf}{2sw}$$

More complex formulae involving several elements have been developed to optimise the shape of the whole building, but although interesting as research tools these result in over-simplification which renders them of little practical help.

Further cost modelling techniques

We have looked at the way in which simple traditional cost models have developed, noted their deficiencies and put forward a list of criteria for judging cost models. It is now intended to look at the possibilities of adopting other models, currently extensively used in other disciplines, in the search for better cost information.

When we refer to better cost information what do we really mean? There are probably five major ways in which it is possible to advance. We can:

• provide cost information *quicker*;
• provide *more* information so that a more informed decision can be made;
• provide *more reliable* cost information which will introduce more assurance into the decision-making process;
• provide information at an *earlier stage in the design process*;
• provide information in a *more understandable form*.

By harnessing the power of modelling techniques it is hoped that each of these objectives can be achieved, or at least a step can have been taken to improve the chances of achieving them.

The traditional models evolved in the way that they did because it was expected that manual labour would be employed to do the calculations. This resulted in an oversimplification of the models, at all levels, for the sake of expediency.

A new view of what is needed in terms of cost information, without reference to the constraints of manual computation, is required if the full potential of the new technology is to be tapped for the benefit of the client. However, almost by definition, all models are simplifications of the thing they seek to represent and are consequently imperfect.

Classification of models

There are a number of ways of classifying models. For example, classification may take place according to the function they perform, e.g. evaluating, descriptive, analytical, optimising, etc. On the other hand it may be according to the form of construction:

- iconic (physical representation of the item under consideration);
- analogue (where one set of properties is chosen to represent the properties of another set, for example, electricity to represent heat flow);
- symbolic (where the components of what is represented and their inter-relationships are given by symbols).

Perhaps the most important knowledge concerning any model is an understanding of its limitations within the context of its use. It is dangerous to ignore the simplifications which are inherent in the construction of cost models in particular.

A considerable amount of research work has been carried out in the universities in developing models during the past 20 years. However, few if any of them have had any impact on the practice of cost planning in the real world. This is almost certainly because the vast expense of developing experimental computer systems into robust, foolproof and fail-safe commercial packages has not been seen as worthwhile in the context of potential use and profitability. Potential users of computer-based cost models are therefore very limited by what is available as commercial packages.

Summing-up

Building cost models are basically of two kinds:

- a product-based cost model is one that models the finished project;
- a process-based cost model is one that models the process of its construction.

This chapter deals with the first of these. The simplest form of such a model takes no account of the configuration, or details of design, of the building but is based simply upon *one* of the following:

- the floor area of the proposed project (gross or net);
- the volume of the proposed project;

- some user parameter such as number of pupil places for a school or number of beds for a hospital.

More complex cost models include the bill of quantities and elemental cost analysis. An element has been defined as 'that part of a building which always performs the same functions irrespective of building type and specification'. The most commonly used list of elements is that published by the RICS as the Standard Form of Cost Analysis (SFCA). There are a number of useful rule-of-thumb generalities about building form and cost which tend to favour low-rise buildings with a low wall/floor area ratio, but there are some exceptions to these.

Further reading

Flanagan, R. & Tate, B. (1997) *Cost Control in Building Design*. Blackwell Science, Oxford.

Chapter 13
Cost Data

Introduction

In reviewing traditional cost models it was suggested that the techniques used are merely the structure around which the professional cost adviser's judgement is applied in order to make them work. The model itself is ill equipped to produce a reliable answer on its own, and reliability at any stage of refinement is entirely dependent upon the costs applied to the measured quantities in their various forms. If the right cost figure is applied to any of the single price methods then they will give better results than more sophisticated techniques with the wrong cost data applied to them.

The factor that makes detailed cost planning techniques more satisfactory than traditional methods is the control that is exercised as the design develops. Even this control, however, is based on unit rates applied to abbreviated quantities (another model) and is therefore dependent on the reliability of cost information.

At the end of the day any estimate is dependent on the prevailing market conditions at the time of tender rather than conditions at the time of making the estimate, and this again requires information in order that market trends can be detected.

So at the root of all this forecasting and control activity we find the need for cost data to supplement the numbers, areas, volumes, etc., which have been used to describe the building. It is this data which is critical in determining whether an estimate is reliable or not.

Problems with computer models

One of the major problems in more recent models, in which computer techniques are used, is the need to ensure that the information on cost held within the memory of the computer system (i.e. the database) is reliable and relevant for all conditions of use of that model.

Very often human intervention is not envisaged in the computer operation, and professional judgement is left until the results have been tabulated. This is very often too late, because the adviser has little chance of knowing why a particular figure has been generated without considerable investigation. And if the adviser is going to adjust a result to what is felt to be more reasonable, then a traditional model, using the adviser's judgement from the start, might as well have been used.

From the earliest days of computers there has been a maxim GIGO ('Garbage In – Garbage Out'), which is just as true with today's sophisticated systems, and this reinforces the view that wrong data used in any technique will produce incorrect results.

Types and origins of cost data

The study of 'what information should we use?' and 'where does it come from?' is essential for the correct understanding of the use and development of costing techniques. First we need to understand what the information is used for.

The uses of cost data

If we were to look at the work being undertaken in a typical quantity surveying office we would find cost information being used for four main purposes:

- The control and monitoring of a contract, for which the contractor has already been selected, through interim and final valuation procedures.
- The estimation of the future cost of a project and the control of its design to ensure that this figure is close to the tender figure.
- The 'balancing' of costs in a cost plan to ensure that money has been spent in accordance with the client's priorities.
- The negotiation of rates with a contractor for the purposes of letting a contract quickly.

Arising out of these activities are other uses which might well be the concern of a special section in a quantity surveying practice or in the research department of a university. These uses relate to a deeper understanding of the external political and economic trends, and the relationship between design decisions and the degree to which they affect cost. These studies usually require a more detailed level of investigation, involving the analysis of large quantities of data or the classification and structuring of the data in a particular way to assist in the development of evaluation models.

To incorporate all uses of data we can probably classify them under four main headings:

- Forecasting of cost
- Comparison of cost
- Balancing of cost
- Analysis of cost trends

Forecasting of cost

Under this heading would come such information as:

- cost per square metre for various types of building;
- elemental unit rates;

- BQ rates;
- all-in unit rates applied to abbreviated quantities.

This information would almost certainly arise from an analysis of past projects, or 'historic costs'. The figures would be updated by the use of a building cost index, itself a form of cost data, and would be projected forward to the proposed tender date by an intuitive or calculated prediction technique.

In recent years the problems associated with historic costs (see later in the chapter) have spurred researchers and practitioners to develop techniques which rely on current resource costs (labour, plant and material) to forecast the cost of a project. This certainly helps in the negotiation of contracts and also in the reliable modelling of the building process. Resources are where the cost is generated, and the adviser is therefore dealing with the origins of cost.

However, the problem lies in the acquisition of information on resource costs by people who are not members of a large building organisation, and in the time-consuming task of synthesising costs from detailed resource inputs.

In fact techniques employing resource costs to forecast building costs have various ways of simplifying the problem to avoid the user being overwhelmed by too much data. These usually involve concentrating on major items and quantities at the expense of those with less cost significance. However, it should be noted that even these techniques rely heavily on 'historic' information regarding the *normal* time and quantity requirements for the resources required for a particular item or building.

Comparison of cost

In this use of data the need is not so much to discover what the building or component will actually cost at the time of tender, but to make a comparison between items with similar function, or buildings of different design, to decide which is the better choice.

The problems of the tender market are not so critical here unless there is evidence to suggest that there may be a change in the cost relationship between the alternatives by the time the decision takes effect.

The criterion in choosing data for this task should be that it is structured in such a form that if the design or specification of an item is changed the use of the cost data will reflect the true change in the cost of the commodity.

Balancing of cost

In determining a budget for the cost control of a building it is necessary to break down the overall cost into smaller units. These smaller units are used not only for checking purposes but also to allow a cost strategy for design to be developed. This strategy will attempt to spend money in accordance with the client's requirements, by allocating sums of money to the various major components of the building.

Data for this use is usually obtained from past projects, and it may be in the form of actual costs or as a proportion of the total cost of a similar project. At the stage that this information is used the design will not have been developed (and may not even have commenced), and therefore the categories used for this breakdown will tend to be broad and probably not exceed 40 items for any one project. The data can therefore be described as 'coarse' as opposed to 'refined'.

Analysis of cost trends

Of paramount importance in any prediction technique which seeks to project costs into the future is information which tells us what is happening to costs in the industry over a period of time. By looking at the way in which costs for different items are changing in relation to one another, or changing between one point in time and another, it is possible to have a better chance of selecting the specification which will suit the client's requirements over the short and long term. It will also allow us to obtain a more reliable prediction of what the market price to the client will be when the job eventually goes out to tender.

These trends may be shown as the change in the cost of materials and labour, or they may be a detected change in the total cost of a particular type of building or component. When data is used for the detection of cost trends the cost is very often related to a base-year cost, in which case the presentation of the information is in the form of a cost index.

It should be noted that the detection of a cost trend does not necessarily imply that it will continue into the future. Indeed, it is very unwise to extrapolate a cost movement without taking into account all the political, social and economic factors that contribute to a change in cost levels – just extending a straight line or a curve without thinking about it is a recipe for trouble.

The number of variables involved in such an evaluation is enormous, and consequently the establishment of what future costs will be is nearly always left to the professional experience of the cost adviser, who will probably take account of economic reports – which often differ widely in their predictions.

It has been said that when you get two different economists discussing a particular economic problem you are likely to get three different opinions!

If these are the major uses of cost data, we can now look at the problems in retrieving and storing the information for these particular applications.

How reliable is cost data?

Nearly all data used in cost planning techniques has been processed in some way, very often with a corresponding loss in accuracy and certainly with some loss of context.

The published information received by the quantity surveyor/cost planner usually relates to a 'typical' building in a 'typical' location with only a brief summary of the contributing factors to such a cost (i.e. the explanatory variables) included. The feedback of such cost information centres in the main around the BQ.

Figure 13.1 shows a diagrammatic representation of traditional cost retrieval and planning practice.

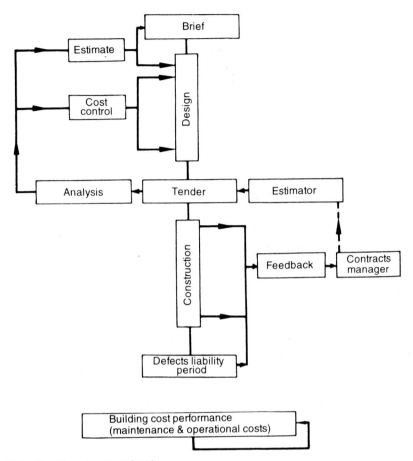

Fig. 13.1 Traditional cost retrieval.

In theory, information on wages, materials and plant is collected on site and fed to the contractor's contracts manager, who then passes this raw data to the estimator. The estimator uses it together with data from other projects to forecast the next tender price. The consultant quantity surveying firm then analyses the contract BQ from the new tender to provide it with data which it then uses to forecast the cost of the next similar building and to control its cost during the development of design. When this new design is being built, the feedback cycle then starts all over again.

The above procedures, although they purport to represent the usual practice, and sound convincing enough, are fraught with problems.

Occupation costs

Notice that in the diagram no feedback is shown into the general system from the area of building performance cost (i.e. maintenance, repair and operational costs). Although some public authorities have attempted to obtain data on these costs and pass this information to the design team for consideration in their choice of detailing and specification this is far from being common practice.

In the private sector the reluctance to collect and divulge occupation costs has been contributed to by:

- the comparatively small number of clients involved in large continuing programmes of building;
- unwillingness to spend the amount of time involved;
- the tax skeletons in the cupboard.

Problems with site feedback

Let us consider the contractor's estimating feedback cycle first of all. The difficulties are as follows (and are further considered in Chapter 18):

- Although records of labour and material costs are kept on site, the information is recorded for the payment of wages and checking of invoices, and for monitoring progress on site as an aid to contract management. These purposes dictate a format which is not compatible with the way in which the estimator works. The grouping of items into 'operations' for reference to the contractor's programme creates difficulties in rearranging the information for the pricing of a BQ based upon the Standard Method of Measurement requirement to measure items as fixed 'in place'. Consequently feedback to the estimator tends to be at the very general level of, for example, 'We were a bit low on plaster rates on this contract Jack' rather than more specific data.

- Even if site data could be related back to the estimator it is unlikely to be of very great benefit, because performance of the labour force varies from day to day according to:
  o weather
  o supervision
  o industrial and personal relations
  o obstruction by other trades
  o the skill with which the work is planned and organised
  o alternatively, lack of clear instruction
  o waiting for instructions on design changes
  o waiting for delivery of materials
  o accidents
  o replacement of defective work
  o failure by sub-contractors
  o psychological pressures.
  It is unlikely that individual performance on one contract will be repeated on another. It will be explained in Chapter 21 why 'standard costing' methods used in industry do not work on a building site. The estimator can only hope to get close to the total labour content of the project and hope that good site management will avoid wastage of this resource.

- No formal record of the performance of labour will be available for work where 'labour-only' sub-contractors are used, and therefore an extremely large area of work is no longer subject to scrutiny and analysis.

- Material use and costs will also be variable between sites, and here again no firm data can be expected for the estimator to use. The amounts used will depend upon such factors as:
  - care in ordering
  - site control and supervision
  - vandalism and site damage
  - replacement of defective work
  - the competence of the workforce.

- Plant costs will be even more difficult than the other categories to relate from one site to another, and depend heavily upon:
  - the quality of site management
  - the amount and type of plant available (especially where the building firm uses its own plant)
  - the extent to which the work lends itself to efficient plant use
  - the amount of disruption to efficient use caused by the work programme, and particularly by changes or delays in this.

The problems with feedback mean that very little data is kept in the contractor's office. The database for the majority of estimators still consists of a 'small black book' of labour and material constants, a list of addresses and telephone/fax numbers of suppliers and sub-contractors, and a good deal of experienced judgement!

No wonder it has been suggested that a tender represents 'the socially acceptable price' rather than a scientific appraisal of the resource needs of the project. Indeed there is some evidence to show that estimators bid randomly within plus or minus 10% of the actual contract cost excluding profit. This is considered to be the limit of their ability to forecast reliably what the job will cost, and is discussed later in this chapter.

Problems with analysis of BQs

The estimator's problems with the retrieval of site data set the scene for the problems the cost planner will find when analysing the priced tender document which the estimator has prepared. The vast majority of the information used by the cost planner arises from the BQ, and yet, as we are beginning to see, this document may be based upon incorrect assumptions which just happen to work in most cases and are convenient for the estimator to use. It has been said that the only reliable figure in the BQ from the client's point of view is the total (i.e the tender figure).

Variation in pricing methods

If we start to break down the fairly similar totals of the different contractors' tenders for the same project into smaller units we are liable to find substantial variation in each individual sub-section when considered in isolation from other sections.

The greater the number of categories the greater the degree of variation, for the following reasons:

- We know that the rates in the BQs are not true costs, but they are not even true prices in the sense of a price on a supermarket shelf. The contractor is not offering to 'sell' brickwork at so many pounds per square metre but to construct a whole building for a total sum – the rates are simply a notional breakdown of the total price for commercial and administrative purposes. There is thus no reason why any individual rate should be justifiable in relation to either cost or competition.

- The way in which the contractor prices the 'preliminary' items of plant, scaffolding, etc., and the firm's own profit and overheads, will vary from one firm to another. Some will place these in the 'preliminaries' section while others will include them, in whole or part, in the measured work rates as they see a commercial advantage in not identifying them too explicitly. Consequently the 'cost' of any individual trade, element or BQ item will vary according to the treatment afforded to these factors.

- It may be in the contractor's best interests to 'load' the prices of those parts of the building which are executed first, such as excavation and earthworks, to improve the project's cash flow (i.e. obtain money earlier in interim valuations to help finance the rest of the work – see Chapter 17).

- In a similar fashion the contractor may anticipate variations to the contract and reduce the price of work which is thought likely to be omitted, while increasing the unit rates of any items which are likely to increase in quantity.

- In addition to the deliberate pricing method of each contractor there are all the variations in unit rates caused by different assumptions being made by each estimator with regard to the resource requirements. These variations tend to be highest in high-risk trades such as excavator and carpenter, and lowest in those most easily controlled such as glazing and concrete work. The assumptions made will relate to the estimator's view of the firm's expertise and economic structure.

- In an estimate for a complex product such as a building it is inevitable that mistakes will be made. These will tend to cancel each other out, but sometimes there is an error of cost significance in a single item which may not be spotted by the quantity surveyor.

If other factors such as the contractor's previous experience of the particular design team or client (e.g factoring-in a premium for a difficult architect or the quantity surveyor!), knowledge (or otherwise) of the locality, experience of the type of building proposed or keenness of tendering are taken into account it can be seen that data obtained from BQs is likely to be highly variable even for contractors who are tendering for the same job.

Variation in BQ rates for different jobs

If we now consider the BQ rates of contractors tendering for different projects then we can expect variability due to all the above factors plus a good few more. These additional items will relate to the site and contract conditions for each particular job, and will include the following factors.

Site conditions

There are a number of site conditions which can affect BQ rates:

- problems of access;
- boundary conditions, especially the problems of adjoining buildings;
- soil bearing capacity and consistency;
- topography and orientation of the building, which may affect manoeuvrability on the site and the type of resources (particularly plant) that can be used.

Design variations

The way the building is designed has an enormous effect on the efficient use of resources, and on production method and time. For example:

- If there is a poor repetition of formwork then it is reasonable to expect higher prices if this fact has been communicated to the estimator.
- Wet trades requiring drying and curing time will possibly create more delay than if a dry form of construction is used.

The extra costs, if any, will be represented in the rates or will result in a redistribution of prices, for example more money in preliminaries for the longer supervision required and plant not fully employed, and perhaps less money in the work sections of the BQ.

Contract conditions

The impact of these is very difficult to anticipate, particularly with regard to their effect on unit rates. In general it can be assumed that the more onerous the terms as far as the contractor is concerned, the more likely it is that a higher tender will be put forward. However, when a contractor is short of work or particularly wants a certain job (say for prestige purposes) then the effect of 'tough' clauses may be less than would normally be expected.

Any contract which requires more working capital for the project, or additional risk, is almost certain to incur a cost penalty which will be passed on to the client. In addition, conditions relating to the length of the construction period, particularly a shortening of time, and the phasing of the works, may both result in uneconomic working and will influence the estimator's rates and particularly the 'preliminary' items

Size of contract

For each contracting firm there is an optimum size of contract that will suit its particular structure and resources. Large firms very often create 'small works' divisions of the main company to deal with those projects which are small or of a specialist nature, such as restoration or fitting-out work, and which cannot carry the overheads of a giant corporation.

Smaller firms, on the other hand, find it difficult to gear themselves up to a multi-million pound project with its specialist plant, complex labour relations and sophisticated supervision and control requirements. The size of the firm in relation to the contract for which it is bidding will therefore affect its approach to the estimate.

The unit rates for a large project should make allowance for the economies of scale that could be expected. However:

- the problems of site working,
- the lack of advanced mechanical production plant and
- the nature of the industry

do not always allow these economies to be made.

Location

This factor will obviously affect the problems of accessibility to the production resources. The transport of materials, workpeople and plant to site, with perhaps accommodation of the workforce on remote sites, makes this an important consideration.

On top of these problems are those of local climate which, even in the UK, may affect the starting on the site, the degree of protection required and the interruption of the work programme.

Despite these problems it should be noted that of all the old principal government cost limits, only the housing cost yardstick identified location as a cost variable and divided the country into regions. The other yardsticks made no allowance for regional variations, but did take the problems of a particular site into account by the use of an 'abnormals' allowance.

A contractor's estimator is likely to pass on the problems of a particular area as an extra cost to the client if the firm has had experience of that location. This again will create variation in the rates due to different allowances being made.

Research into variability

All the problems listed contribute to the wide variability in BQ rates.

Derek Beeston of the former Property Services Agency looked at the major items in each work section (or trade) in a large sample of BQs from different government projects and compared the unit rates for each individual item. He found that the variability tended to be different for each trade, and expressed the degree of variability in the form of the 'coefficient of variation', which enables a comparison to be made between the variability of items of different value and is given by the formula:

$$\text{Coefficient of variation} = \frac{\text{standard deviation}}{\text{arithmetic mean}} \times 100\%$$

So in concrete work, for example, the cost per cubic metre may be £33.00 and the standard deviation £5.00:

$$\text{Coefficient of variation} = (5/33) \times 100\% = 15.15\% \text{ say } 15\%$$

This can now be compared with plaster work with a cost of £3.00 per square metre and a standard deviation of £0.30:

$$\text{Coefficient of variation} = (30/300) \times 100\% = 10.00\%$$

The standard deviation on its own would not allow the degree of variability in each trade to be compared.

The results of Beeston's investigations showed the following:

Excavator	45%	Joiner	28%
Drainlayer	29%	Roofer	24%
Concretor	15%	Plumber	23%
Steelworker	19%	Painter	22%
Bricklayer	26%	Glazier	13%
Carpenter	31%	All trades	22%

It is no surprise that excavator, with its high risk problems due to weather, soil conditions, accessibility, and so on, should have the highest variability. The ranking of some of the others is perhaps unexpected and a detailed investigation would need to be undertaken to discover the reasons for their particular performance.

However, the table does illustrate very clearly the problems of using BQ data. Once again, success depends upon the skill of the user in determining what rate to use against a particular measured quantity. A computer cannot make this judgement.

The contractor's bid

If we accept the problems associated with obtaining reliable feedback from site and the rather arbitrary value given to unit rates in the BQ then we might wonder how the contractor arrives at a 'bid' for a particular contract when he is in competition with others. The theory is that the estimator:

- *knows* how long it will take a team of craft operatives and labourers to execute a job;

- *knows* the precise amount of material required including wastage;
- *can define* the exact output of each item of plant to be used and the time it will be on site;
- can add on the *measured* cost of supervision and overheads;

to arrive at the net cost of the job excluding profit.

The process is essentially *deterministic*; it assumes that the estimator (if clever enough) can determine the exact amount of resources and their cost. However, every contractor knows that this is not possible.

The notion of non-deterministic pricing

Research has suggested that while the above process is theoretically adopted it is in fact very often used to justify a total that the contractor has in mind from the very beginning. Brian Fine, who used to work for one of the largest UK building companies and later became a management consultant, has suggested that contractors may bid at what they consider to be a 'socially acceptable price'. This is not well defined, but is considered to be the price which society is known to be prepared to pay for a particular type of building. For commercial projects this would probably be related to the income of the office block or output of the industrial plant. For social projects it may be related to the known funds available. In the case of government funded projects it may be a defined cost yardstick.

It is considered that this value of the building is relatively easily calculated and that generally the building is buildable at around the socially acceptable price. By bidding at this value the estimator overcomes the problem of being unable to predict resource requirements for the project. As the client's anticipated cost is probably based on what has been successfully built before it is therefore likely that the estimator will have put forward a reasonable estimate, and the labour and material 'constants' are used to justify this.

It is unlikely that many contractors consciously go through the above process, and indeed they often react strongly to the suggestion that they do. However, despite the wide variation in individual BQ rates the final tender figures are often considerably closer than one would expect. A coefficient of variation for the spread of tenders on a project tends to be in the range of 5–8% and this can be applied to a wide variety of projects.

Support for the notion of non-deterministic pricing

Contractors will occasionally admit that they can write down the cost of the job before they start pricing it, assisted by:

- Working out the yardstick amount before pricing a government job.
- The client naming the approximate value of the contract when enquiring if the contractor is willing to tender.

- The building grapevine of sub-contractors and suppliers, who probably quote several times for the same job but for different contractors, and ensure that the anticipated contract sum is well circulated.

The case is further strengthened by the following:

- *The name of the project appears to influence the cost.* A good example of this phenomenon was the decision by a number of hospital boards some years ago to let nurses' accommodation blocks as a separate contract to the main hospital development. The reason for this decision was that the nurses' accommodation was costing appreciably more than comparable students' accommodation in the education sector. Tenderers had apparently applied the same criteria of complexity, difficulty and high cost associated with hospital projects to the far simpler problem of the nurses' accommodation blocks. Hence of course the high prices, and the suggestion that the real cost was not being estimated.

- *Constants are not reliable.* As stated earlier in this chapter, it is not possible to establish accurate constants of labour, plant and material to apply to the job that is being estimated.

  A typical output of a gang of bricklayers on site may be as shown in Fig. 13.2. Contrary to what is often asserted, data of this nature is unlikely to be very helpful in estimating the cost of the next job.

  On the graph in Fig. 13.2, week 6 results in a negative output, because the gang had to pull down something they had constructed the previous week because it was unsatisfactory or the subject of a variation order. This is not an uncommon experience, but will it happen on the next job? If confident

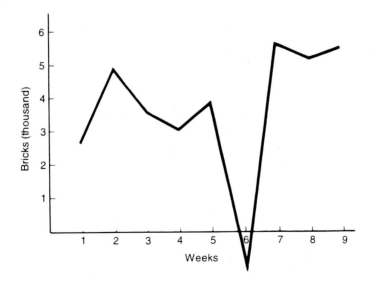

Fig. 13.2 Weekly output of bricklaying gang.

predictions cannot be made then an intuitive answer in the form of the socially acceptable price is perhaps the best answer.

- *Chance of contracts being won.* In a competitive market it appears that contractors take their place in the bidding order by chance. As would be expected with chance, in the long run they appear to win contracts in direct proportion to the number of other contractors they are bidding against.

  J.C. Pimm of Management Procedures (Building) Ltd looked at the number of contracts that four companies had bid for and the number of competitors identified by the total number of bids. From this information, and knowing the number of contracts actually won, he produced the results shown in Table 6. It can be seen that there is a very high correlation between the number of contracts actually won and the number expected to be won by chance, assuming each firm took its place randomly in the bidding list. This is supported by Fine, who suggests that contractors can only estimate within plus or minus 10% of the actual cost. If this is so then each contractor needs to add on a mark-up to the estimate, to allow for the fact that the firm that wins the contract is likely to have made the largest error. If this is not allowed for then the error will come out of the expected profit, or end up as a loss. The amount to be added should depend upon the numbers of contractors bidding.

  In practice, very few firms consciously allow for this factor, but they probably note its eventual impact upon the profitability of a project.

Table 6 Bids – actual results compared to chance.

Company	No. of contracts won	No. of bids	Bids per contract	Per cent if won by chance	Per cent actually won
A	41	249	6.1	16.4	17.0
B	36	183	7.0	14.2	15.4
C	19	88	4.6	21.6	21.1
D	35	202	5.8	17.3	17.1

- *Estimates are tailored to suit the market.* If the estimation of cost really was a deterministic exercise then the constants used in the estimate would be absolute and unchanging unless, of course, they were amended as the result of ascertained cost feedback. However, what happens when an estimator or firm loses a number of bids in succession? The estimator is told to 'sharpen your pencil', and the firm reduces their rates! In order to survive they must get work.

  This practice suggests that the estimate does not mean very much in terms of an accurate prediction. If the firm does get the job and they think it will be 'tight' they put their best management team on the project and very often get a better return than on a job where they foresaw a large profit and did not maintain such tight control! The lower price that they are bidding could be the socially acceptable price.

It would not be fair to suggest that the concept of socially acceptable price has been thoroughly substantiated on the basis of the few arguments considered. However, doubt has been cast on the deterministic estimating process and this in turn should influence the way in which we look at and use cost data.

Obviously the factors of supply and demand related to the capacity of the industry, and its workload, will influence the level of the contractor's bid, and the relationship between these factors and the socially acceptable price has not yet been well defined.

However, it is important to realise that market conditions reflect the economic standing of the country and this aspect may well have more impact on 'historic' cost data, and the price society is prepared to pay for its buildings, than any other.

The structuring of cost data

In addition to the difficulty of acquiring reliable cost data there is the problem of what to do with it once acquired.

The need for a system of cost planning to have access to cost data at a number of different levels of complexity postulates a hierarchy of cost groupings, or even a series of interlocking hierarchies, because some of the criteria to which cost data will be applied are of different types and not merely different scales.

One such hierarchy, which we encountered in Chapter 12, is based upon the concept of *design cost elements*. This is a highly pragmatic breakdown of a project for cost planning purposes which, apart from its inherent faults, suffers from the disadvantage that it does not meet the requirements of any other functions in the design/build process.

Another hierarchy is the *Standard Method of Measurement* format used for BQs, which is only of limited use for other purposes. However, because the information provided in the BQ is:

- generally of a high quality,
- tied back into the contract,
- available to the builder without further cost or trouble, and
- structured in accordance with the 'Common Arrangement of Work Sections for Building Works' produced by the Co-ordinating Committee for Project Information (CPI),

its format does tend to pervade documentation produced for process purposes rather more than might be expected.

Operational cost centres, which are useful for actual control on site, are difficult to relate to the estimating and commercial systems of the industry, except at the highest levels of generality. Because of the obstruction to communication between builder and designer which is caused by the price mechanism and the associated contractual standpoints there has been little attempt to link site operation costs to design criteria.

An integrated system of groupings?

A number of efforts have been made, over the last 25 years in particular, to devise an integrated system of groupings of building work which would meet the needs of all participants in the project, from the point of view of:

- design process;
- costing;
- communication;
- organisation;
- feedback.

It is argued that such a system would reduce duplication of effort (it is alleged that the same piece of building work may be physically measured up to 15 times for various purposes), improve communication between the parties and permit a far higher quality of costing and cost feedback.

The CI/SfB system whose use was encouraged by the RIBA was one such effort, but was too oriented towards its original purpose of filing design information to have found much favour elsewhere (it is now incorporated in the wider *Uniclass* system).

A much more complex version of the same basic system, CBC, was developed in Scandinavia and relied heavily on computer use at a time (the 1970s) when this imposed considerable constraints on a system. In Britain the then Department of the Environment set up a data co-ordination study with substantial resources and a large measure of contribution from industry and the professions, it being their very reasonable view that this was the only way of obtaining results which would be useful, and acceptable, to all.

Unfortunately the work ceased, for party political reasons, without producing much more than a number of very interesting reports and the *Construction Industry Thesaurus*, which is used mainly for library, information and reference purposes. Little further work has been undertaken in this field in more recent years and little progress has been made, but there is plenty of (rather dated) background reading available.

Sources of data

So far in this chapter we have dealt with some of the more general problems of cost information associated with its use, retrieval and structure. Although these problems should be identified it has to be recognised that the quantity surveying profession has been using data obtained from BQs for scores of years with reasonable success. Contractors, too, have been able to prepare realistic estimates from the very generalised feedback that they get from site.

The problems associated with data should not therefore be allowed to obscure the fact that present data sources available to the industry have allowed the development of a simple and reasonably effective system of cost forecasting and control that the various parties are familiar with.

From the point of view of the client's cost adviser, information can be obtained from two main sources:

- data passing through the cost adviser's office;
- published material found in the technical press, price books and information systems.

There is no doubt that most cost advisers would normally prefer to use their own data, for the following reasons:

- The data will be from a project whose background they know. They will therefore be aware of all the problems associated with the project:
  - its location
  - the market conditions
  - complexity, etc.

  which influenced its price, and they will also know about the project's outcome.

  We have already said that traditional cost modelling requires sound professional judgement. This is better served by using information which is known to the cost planning organisation through involvement in the project rather than received second-hand from a published source, in an abbreviated form and with little background information.

- The further detailed breakdown of any structured information is available should it be required. To take the example of an elemental cost analysis, the firm will have available:
  - the full priced BQs
  - the original measurements
  - the working drawings
  - the contract

  should they wish to find further information on why an element cost a certain amount or what were the prevailing market conditions at the time and whether the overall price was considered high or low. It is very unlikely that this level of detail would be available in any published data, however well prepared.

  To take another example, many quantity surveyors/cost planners prepare their own cost indices because they know the weightings and costs included and these can be manipulated with greater confidence in assessing the needs of a particular project.

- The data will refer to the geographical area in which the firm carries out most of its work.

- Compared to published data there is a shorter time lag between receiving raw data, processing it and being able to use the structured information in the office. This is particularly relevant to cost indices, where the published information may be anything up to 6 months behind the times.

- The choice of classification and structure of the processed data is under the control of the firm. It is therefore possible to ensure that the details which are most relevant to that particular practice are emphasised and identified rather than having to accept a standardised published format. If errors and ambiguities occur it is also easier to spot them in your own system rather than somebody else's.

Having established the preference of most cost planners for their own information it would be unrealistic to suppose that any but the largest firms or organisations could exist solely upon the data passing through their office.

The range of projects undertaken by most practices is considerable, varying from government and local authority buildings to commercial and leisure buildings for individuals and corporate clients. The variety of specification, location, size and shape of this range is also large, and it is therefore most unlikely that the office will have up-to-date information which it can apply to all its new projects.

Published cost data

The common unit rates which each quantity surveyor will gather from handling project information will help in preparing an approximate quantities estimate, but at the coarser level of elements and single price estimates a wide enough variety of data is unlikely to be readily available.

While the cost planner can probably make intuitive assessments, particularly of cost per square metre, it is reassuring to have a check available in the form of published information. Indeed, it is probably the primary role of published information to provide data which enables practitioners to check on their own knowledge and to provide a context for their own decision-making.

The following are the major sources of this kind of data, which apply almost entirely to conditions in the UK. Such information as exists overseas is often of a more condensed statistical nature and makes the cost planner's task in most other countries more difficult.

The technical press

There has been a steady improvement in the quantity and quality of cost information appearing in the journals and magazines concerned with design and building management.

Until some 30 years ago almost no information was disseminated in the press regarding costs of buildings. This was always jealously guarded as confidential until *The Architects' Journal* took the lead in refusing to publish accounts of new buildings unless cost information was included.

Since then, however, largely due to demand from clients for better cost control, editors have found it necessary to devote more space to cost analyses and forecasts. The type of information that can now be found in such journals as *The Architects' Journal* and (particularly) *Building* magazine includes:

- elemental analyses;
- resource costs and measured rates;
- cost indices with future cost projections;
- analysis of the economic performance of the industry;
- cost studies of different types of buildings.

Indeed, with the rapid changes in the level of resource costs and measured rates over very short time-spans the technical journals, published at weekly intervals, have largely replaced the traditional annual standard price books as a source for quick reference material.

Builders' price books

A traditional source of useful information has been the price books which are normally published annually by several long established organisations. These too have expanded the scope of their contents to keep pace with current demands and to try and maintain an edge over the technical journals. However, in periods of rapid inflation or market changes they have suffered the drawback of having to prepare their information well in advance of the year of publication. This time lag of several months, and the 12-month period for which the information is meant to be current, means that the rates have tended to be forecasts of what is likely to happen at some unknown point of time in the year ahead.

This problem has more recently been overcome by the publishers offering on-line computer access to updated information.

Despite the above problems it is usual to find an assortment of *Spon's Architects' and Builders' Price Book*, *Laxton's Price Book*, *Wessex Price Book* and *Hutchins Priced Schedules* on the shelves of all the building professions. Their popularity stems from the very comprehensive nature of the data now included, and subsections normally cover:

- professional fees for building work;
- wage rates in the industry;
- market prices of materials;
- constants of labour and materials for unit rates;
- unit prices for measured work in accordance with the Standard Method of Measurement;
- elemental rates for a variety of work;
- building cost index;
- approximate estimate rates and comparisons;
- cost limits and allowances for public sector buildings;
- European prices and information.

Spon also publish separately a very comprehensive *Mechanical and Electrical Services Price Book*. They also publish a *European Construction Costs Handbook* and an *Asia-Pacific Costs Handbook* for overseas work.

As with data from any source, considerable care needs to be exercised in the use of the information in all these books. For example, in the Prices for Measured

Works – Major Works section of the 1999 edition of *Spon's Architects' and Builders' Price Book* the prices are for a contract of about £1,300,000 in value (excluding preliminaries), in the outer London area. They assume that a reasonable quantity of work for all items is required. The subletting of all work which is normally sublet (a builder's price for this type of work is likely to be higher than normal market rates) is also assumed. No allowance is included for overhead charges and profit, nor for preliminary items or VAT. The prices are based upon a tender price index of 325 in relation to a base of 100 in 1976 (which may not be the same as the index normally used by the cost planner).

The user is expected to make all allowances for changes in the location, size of contract, small quantities of particular items and market changes.

It will be seen from this how dangerous it is to start picking out prices at random from various publications, all of whose rules may be different, without reading the detailed conditions in each case.

However, price books have a very useful role to play in enabling practitioners to check on their own knowledge and assumptions.

Information services

With the advent of cost planning techniques and the need for a wide range of information on different types of project it was realised that most offices would require information additional to that normally found in one particular practice.

It was suggested that if firms could be persuaded to supply their own elemental cost analyses to a central body, then a large pool of information could be gathered which could then be disseminated to the subscribing members. If these analyses could be supplemented with other more general information on trends and economic indicators, then a useful service would be provided to the profession.

Therefore, in 1962, the RICS set up the Building Cost Information Service (BCIS) to undertake this task. For its first 10 years the service was only available to QS members of the RICS, but since 1972 it has been available outside the quantity surveying profession.

Today there are other cost information services, but the breadth of information and independence of the BCIS, and its sister service Building Maintenance Information, makes these RICS services the leaders in their respective fields. They are dealt with here in some detail.

Building Cost Information Service

BCIS is a subscriber-based service which collates and analyses data submitted by its subscribers and incorporates material from other sources. The information is interpreted by the BCIS professional staff of chartered surveyors, and is presented in two formats:

- BCIS Bulletin Service – a hard-copy service;
- BCIS *Online* – an electronic data service.

Naturally, all information provided by subscribers is treated in confidence. Approval to publish project-specific data, such as elemental analyses, is sought from subscribers and any other parties to the information.

A summary of the BCIS Bulletin Service, which principally deals with capital cost data, is given below.

Indices and forecasts

Published quarterly, this is a bound report including:

- the complete series of BCIS Indices;
- a 24-month forecast of the main tender and cost index series;
- an executive summary;
- a thoroughly researched commentary on market conditions with trends projected over the coming 24-month period;
- a range of individual indices for tender prices, regional prices, input costs and output prices.

Surveys of tender prices

A quarterly bound publication, the principal content of which is a range of current pricing studies incorporating updates of:

- the BCIS Tender Price Index;
- the BCIS Building Cost Index.

This publication provides:

- information on average building prices for different building types, new and refurbishment, expressed in £/m^2 of gross internal floor area and adjusted to current prices using the BCIS Indices;
- functional prices, such as £/pupil place for schools, for around 200 building types;
- studies of price differentials by size of contract, location, and type of work;
- results of a survey of the percentage added for preliminaries, dayworks and sub-contract work.

It should be noted that the BCIS Indices are based on an index of 100 in 1985, and so are quite different to those used by *Spon's*, for example. Previous warnings about care when combining information from different sources should be remembered.

Special studies

BCIS publishes special reports periodically throughout the year. Currently there are four annual special reports:

- *The BCIS Histogram Study*. This study is intended to complement the £/m^2 prices published in the *BCIS Surveys of Tender Prices*. The Histogram Study presents the distribution of costs in graph form and shows the variation in prices which can arise for the same functional type.

- *BCIS Project Prices*. This is a supplement to the *BCIS Surveys of Tender Prices*. It gives:
  - details of around 600 individual projects submitted by BCIS subscribers over the previous 2 years;
  - £/m^2 data on the most recent projects submitted, grouped by function, so as to give an indication of current tender prices;
  - £/m^2 at 1995 constant prices to facilitate easier comparison of project prices;
  - a summary of recent projects for subscribers, identifying where further information is available.

- *BCIS Average Group Element Prices*. This gives average prices, and a distribution of prices (lowest, 10%, 25%, 50% median, 75%, 90%, highest), for each of the five BCIS Group Elements on each of a range of building types.

  These are not carried forward to totals for each building type, as the sums of the lowest prices and of the highest prices would be well outside the actual range of costs in the Project Prices Study.

- *BCIS Five-Year Forecast*. This was first published in July 1997 in response to requests for forecasts beyond the year 2000 from subscribers involved in Lottery and Millennium bids. The report contains a commentary on the background to the forecast including an examination of medium- and long-term trends in both the macro and construction economies. Forecast indices are provided for:
  - tender prices;
  - market conditions;
  - building labour and materials costs.

Elemental analyses

These are a key source of price information from accepted tenders and are published quarterly as loose-leaf data sheets classified for filing by building type. There are two levels of analysis:

- *concise elements* contain £/m^2 for six group elements;
- *detailed analyses* break down tender prices into £/m^2 for 34 elements.

The lists of elements are as set out in Chapter 12.

Labour hours and wages and dayworks

These are published periodically as loose-leaf data sheets, in which details are given of the principal rates in the building and allied industries. The information shows:

- rates and operative dates for current and recent settlements, along with the dates they were promulgated by the appropriate wage-fixing body;
- amounts of current National Insurance contributions, holidays-with-pay and CITB levy;
- daywork rates (standard hourly base rates for use in dayworks) calculated by BCIS for each change in wages, National Insurance contributions and other variables.

News

BCIS News accompanies each *BCIS Bulletin*. It takes the form of a newsletter highlighting items contained in the latest bulletin. It also summarises items of economic significance and reports on BCIS plans and activities.

Digests

Digests of published articles give the source of publication and a brief summary of the subject covered. Construction economics is a broad subject area, and the Digests help subscribers to keep abreast of current literature. A photocopying service is available to subscribers for personal research.

The BCIS also publishes the following documents, which are not part of the Bulletin Service:

Guide to Daywork Rates and Updating Service

This annual guide covers the daywork rates for 54 grades of operative, giving both current and historical rates. Each rate is calculated in accordance with the BCIS/RICS interpretation of the appropriate 'Definition of the Prime Cost of Dayworks'.

Daywork rates can change at any time in the year, so BCIS offers an updating service which informs subscribers of changes in rates as and when they occur.

BCIS Quarterly Review of Building Prices

This contains a selection of BCIS data which gives guidelines on the general level of building prices. It also has a brief commentary on:

- market conditions;
- tender prices;
- location factors;
- average £/m^2 building prices.

Guide to House Rebuilding Costs and Index

This annual guide is widely used by professionals when assessing rebuilding costs for insurance reinstatement cost assessments. It contains the main tables of cost and:

- four regional groups;
- five house types;
- three sizes of house;
- three quality specifications.

The *ABI/BCIS House Rebuilding Cost Index* is published monthly to facilitate index linking of the figures in the Guide throughout the year.

Building Maintenance Information

Since 1971 BCIS have run a sister service dedicated to maintenance and occupancy costs rather than capital cost. Building Maintenance Information (BMI) provides cost and management information about property occupancy and buildings in use.

BMI is available as a subscription service and includes:

- *Quarterly Cost Briefing.* This contains a brief commentary on current and forecast trends in maintenance costs together with cost indices for:
  - redecoration;
  - fabric and services maintenance;
  - cleaning;
  - energy.

- *Property Occupancy Cost Analyses.* These report annual expenditure on individual buildings in a standard (BMI Property Occupancy Costs Analysis) format which includes:
  - decorations;
  - fabric maintenance;
  - services maintenance;
  - cleaning;
  - utilities;
  - administration costs and overheads.

- *BMI Price Book.* This book is dedicated exclusively to maintenance work and contains:
  - labour constants;
  - current prices;
  - item rates.
  It is used by some organisations to form the basis of a bespoke schedule of rates for use on measured term contracts.

- *BMI Information Guide for Facilities Managers.* This provides points of contact for further information and helps users to find appropriate organisations who hold information on:
  - firms who can provide specialist services;
  - manufacturers and suppliers of specialist materials and equipment;
  - technical guidance.

- *Special Reports.* These are reports on maintenance and occupancy costs for

individual buildings, and studies giving averages for a variety of building types. These are intended to help with:
○ benchmarking;
○ planning;
○ forecasting;
○ whole-life costing.

- *News*. *BMI News* summarises the latest industry developments and contains statistical and editorial articles designed to keep subscribers well-informed.

- *Digests*. These provide a brief summary of published articles with a bearing on property occupancy from a wide cross-section of the building maintenance and facilities management press. It is backed by a photocopying service.

BCIS Online

Since 1984 the BCIS has been operating a computer-based service which gives subscribers unlimited access to the entire BCIS databank from their own PC. It can be accessed from anywhere in the world.

With a link to the BCIS host computer a simple menu-driven structure (which was summarised in Chapter 2) enables the user to select the type of data and the specific information required in the shortest time. The system is straightforward to use.

Simple but powerful selection tools cut out time-consuming browsing to locate, for example, the cost analyses needed for a project. This ensures that users have access to the widest possible range of examples as well as the most up-to-date information – particularly important when the market is volatile.

Once the required information has been located users can copy it to their own machine for further use. The system allows analyses to be updated to current or projected pricing levels and adjusted for location. Data can then be:

- printed;
- displayed on screen;
- used in conjunction with approximate estimating and other software such as spreadsheets.

The BCIS Approximate Estimating Package enables users to manipulate cost analyses to produce a budget estimate and cost plan. The Package is designed to replicate the process that a QS would follow when working by hand, but can be amended if desired.

All calculations can be viewed on screen, and the end results can be edited, over-ridden or revised as required. When the estimate is produced the user can then revise individual elements in order to test a variety of 'What if?' scenarios.

Government literature

With the fragmentation of central and local government activities as a result of privatisation the provision of information has been subjected to many changes, and only the preparation of cost indices remains in the Construction Directorate of the Department of the Environment, Transport and the Regions (DETR).

Technical information systems

For many years office library systems which contain a collection of current trade literature relating to building products have been available to the building professions. Their major purpose is to provide a ready reference for the practitioner on specification and performance of a wide variety of building products.

At one time it was not uncommon to find current price lists included, but prices change more frequently than specifications, and the problem of updating this secondary information has been solved by leaving it out. The use of such services for costing purposes, admittedly never their major role, has therefore died out.

Other sources of cost data

While the two main sources of data for the cost planner are the firm's own data and published data, there is a further source available for some types of work, which in some ways is a combination of the two. This is the obtaining of information from specialist sub-contractors and specialist consultants, and with an ever larger proportion of building cost represented by specialist firms undertaking:

- roofing,
- flooring,
- windows,
- doors,
- cladding,
- finishes,
- structural systems,
- landscaping and
- engineering services,

this is an increasingly valuable source.

Like published information it suffers from remoteness, but because there is usually an element of personal contact involved it is often possible to explain and discuss exactly what the circumstances and requirements are. Personal knowledge and contact is also helpful because there is little other guarantee of the soundness of such advice.

If prices are obtained from specialist sub-contractors the cost planner must remember to allow in addition for:

- the builder's profit;
- the builder's contractual discount allowance if the sub-contractor's price is net;
- facilities which the builder must provide including:

- ○ scaffolding;
- ○ unloading and moving materials;
- ○ storage;
- ○ office accommodation;
- ○ use of builder's plant and equipment;
- ○ incidental builder's work in cutting holes, etc.;
- ○ assistance with site fixing. Some firms send only a specialist fixer who does little more than supervise the builder's own site workers.

It is important to find out exactly what the sub-contractor requires in these matters – some such firms are almost entirely self-sufficient, others lean upon the builder as much as possible.

Future development

Reliable cost data is a vital primary need of the cost adviser, and one which increases in importance as modelling techniques develop which require different information structures.

Development generally will almost certainly be in three directions:

- representing the way that costs are incurred on site, with the need for improved feedback;
- presentation of these resource costs in such a way that they can be related to design development;
- use of information technology.

Summing-up

An attempt has been made in this chapter to outline the sources of cost data available to the cost planner and their use and problems.

The primary source of such data is still the priced BQ, but the reliability of this data is rather questionable and the extent to which it even tries to represent actual costs is open to doubt.

Do not forget that the price data included in BQs is confidential and the contractor's permission must be sought if any analysis, etc. is to be published. Builders' own cost systems are not usually as suitable for design cost planning purposes as might be expected. Published cost data of various kinds is a useful back-up. BCIS and other information services are also very useful.

Chapter 14
Cost Indices

The cost index

One of the most important items of cost data, particularly in connection with forecasting techniques which rely on historic (i.e. past) information, is a cost index. Its object is to measure changes in the cost of an item or group of items from one point in time to another.

The inflation index produced by the government every month is a well known example. In compiling an index a base date is chosen and the cost at that date is usually given the value of 100, all future increases or decreases being related to this figure.

Example

Suppose the cost of employing a labourer on site at base date was £160.00 per week and the current figure is £240.00 per week. £160.00 would be given the value 100 and £240.00 would be represented by the figure 150, derived in the following manner:

$$\frac{(240 - 160) \times 100}{160} + 100 = \text{cost index in relation to base } 100 = 150$$

As the base is 100 the number of points of any subsequent index above 100 (in this case 50) also represents the percentage increase since the base date.

However, if we are comparing costs over time in which the data we wish to update is not at base year cost, but has an index of say 120, then to arrive at the percentage increase in cost where the current index is 150 the following calculation is used.

$$\frac{150 - 120}{120} \times 100 = 25\%$$

Notice that the answer is *not* the difference in the number of points on the index scale, i.e. 30.

Use of index numbers

There are a number of uses to which index numbers can be put in the construction industry.

Updating elemental cost analyses

This is perhaps the most common use for the quantity surveyor and is an essential part of the elemental cost planning process. Tender information of past projects can be brought up to current costs for budgeting purposes. However, great care is required when updating information beyond a period of, say, 2 years.

Updating for research

It is extremely difficult to obtain large quantities of cost data relating to the same point in time in order to analyse trends and patterns for cost research. By bringing cost information, obtained at a number of different points in time, to a common date by the use of an index, a much larger sample of data can be examined.

Extrapolation of existing trends

By plotting the pattern of costs measured by an index it may be possible to extrapolate a trend into the future. However, there are very great dangers in extrapolation. For example, during the 1960s there was a steady increase in cost of about 5% per year. Extrapolations were undertaken using this figure for projects 1 or 2 years ahead. This worked quite well for a number of years until 1970, when the cost of building to the client suddenly rose by about 10%. Many quantity surveyors (and other professions) had not foreseen this rise and were heavily under-valued in their budgets.

It may be thought that inflation is fairly constant in the UK at present – but then that is what people thought in the 1960s. Extrapolation based purely on present-day trends is very risky.

Calculation of price fluctuations

During periods of rapid inflation it was the custom for building contracts to be entered into on a basis whereby the contract amount was adjusted during the progress of the work to take account of inflation since the date of tender.

By applying an index to the cost of work undertaken during a specified period, it was possible to evaluate the increase in costs of resources to the contractor more speedily and with less ambiguity than by using the previous laborious method of comparing every invoice price with the base date cost of the same item. The NEDO index (see later) was most often used for this purpose.

Identification of changes in cost relationships

If a cost index is prepared for the different components of a building, or for alternative possible solutions to a design problem (for example steel versus reinforced concrete frame), then it is possible to see the changes in the relationship between one component and another over time. It may then be possible to identify when one solution appears to be a better proposition than another.

Assessment of market conditions

Quantity surveyors are particularly interested in the price their clients have to pay for a building. If the index will measure the market price, as opposed to the change in the cost of resources, then a measure of current market conditions can be obtained which is of enormous benefit in updating and forecasting cost.

Approaches to constructing an index

Over the last two or three decades there have been a number of attempts at providing a reliable cost index. They include the following:

The use of a notional bill

In this method a typical (or synthesised) BQ is chosen and repriced at regular intervals by a competent quantity surveyor based on experience of current rates. Unfortunately, rates vary so much that it is extremely difficult to assess the current tender situation in such a way that a reliable index can be obtained from the resultant totals.

This method is now to all intents and purposes obsolete.

Semi-intuitive assessment

This method is based upon trends in resource costs combined with tender reports by quantity surveyors. By receiving reports on current tenders from each of the project quantity surveyors in their organisations, experienced practitioners can adjust their knowledge of the changes in material and labour costs to prepare an index for the current market situation.

This works quite well where the market is stable or steadily changing, but these judgements are extremely difficult to make at a time when the economy is subject to 'stop-go' conditions.

This method has been superseded by the tender-based index described later.

Analysis of unit price for buildings of similar function

If data could be found giving, say, the cost per square metre of all schools of a certain type being built in the country at one point in time, then the average of these costs could be used to provide the basis of an index. If the exercise were repeated at regular intervals for that type of building then a regular index could be established.

Problems arise, however, due to such large design variables as specification and shape. Such an index would also be only applicable to buildings which are homogeneous in function and standard and for which there is a regular building programme to provide the data.

Factor cost index

If a typical building is analysed into its constituent resources and the cost of each resource is monitored over time, then a combined average index can be prepared which measures the change in the total cost of the building over the same period.

Each resource (labour, plant and material) would need to be given due importance in the index according to its value in the total building. The construction of, and problems associated with, this type of index are described in more detail below.

Tender-based index

There is a great need for quantity surveyors to have a measure of the market price their clients have to pay for their buildings. The accepted tender figure based on the pricing of a BQ is a record of the market price for a particular building at a specific point in time. If the measured items in the BQ are repriced using a standard schedule of base year prices (to give a base year 'tender'), then an index can be constructed by comparing the current tender figure with the new derived base year total.

Short cut methods are available to avoid repricing the whole BQ. Details of this method are provided later in the chapter.

The last two of these approaches now need to be considered in more detail.

The factor cost index

It has already been explained that this uses changes in the cost of resources to build up a composite index.

To illustrate the method let us look at the construction of a simple index for the cost of a brick wall. We will assume that the following table represents the cost of the constituents at base year level in the first column, and then at today's date in the second column. The final column represents the index resulting from these figures (calculated as described previously).

Cost per square metre of wall

	Base year £	Current £	Index
Bricks	20	28	140
Mortar	2	3	150
Labour	8	10	125
			415

Average index = 415/3 = 138.34

However, in the above example no account has been taken of the fact that bricks as a proportion of total costs are ten times more important than mortar, all the resources having been given equal status. Consequently, a very rapid rise in the cost of mortar would have a disproportionate effect on the composite index, and this would not be representative of the increase in the cost of a brick wall as experienced by the contractor.

To overcome this problem the resources need to be 'weighted' in accordance with their importance as follows:

	Index	Base year weighting	Extension
Bricks	140	20	2,800
Mortar	150	2	300
Labour	125	8	1,000
		30	4,100

Average weighted index = 4,100/30 = 136.67

The effect of the weighting is to reduce the index value in this example because mortar (which has increased most in price) does not now share equal status with bricks and labour.

A further problem arises at this point. Suppose for some reason the labour force becomes more productive over the time period, by say 25%. Then at base year costs the value of the labour would have been £6.00 instead of £8.00. This will affect the amount of labour involved in building a brick wall and consequently its cost. This change, although affecting the cost to the contractor and possibly the client, will not be represented in the index unless the weightings are changed.

	Index	Current year weighting	Extension
Bricks	140	20	2,800
Mortar	150	2	300
Labour	125	6	750
		28	3,850

Average revised weighted index = 3,850/28 = 137.5

This is a very real difficulty with the factor cost index because it is not easy to judge productivity over time, particularly with regard to the construction of whole buildings. In fact in the majority of cases, weightings in building factor cost indices are assumed to remain constant until a revision of the index is undertaken and a further analysis of the importance of each resource is made.

Where an index uses the base year weightings for each calculation then it is known as a Laspeyres index. Where the index uses weightings obtained from the current year or point in time that is under consideration then it is known as a Paasche index. Both names derive from their original authors.

Another factor not considered in the above examples is the contractor's profit and overheads. Profit, particularly, can be a function of market conditions and it would be difficult to devise a reliable quantitative measure for this variable. This is why a number of factor cost indices have incorporated an additional component, called a 'market conditions' allowance, very often based on the professional judgement of its author.

Constructing a factor cost index for a complete building

The brick wall was chosen to illustrate the essential ingredients and problems of an index of this type. When constructing an index for a complete building the task becomes more complex although the same principles apply.

A typical procedure can be summarised as follows:

- A typical building (or group of buildings) is selected for analysis into its constituent proportions of labour, plant and material.

- Analysis takes place and the building resources are allocated under various headings to suit the representative cost factors for which information is available. For example, if use was being made of published information for material indices (such as those prepared by the DETR) then the different materials in the building would be analysed according to the structure of that published data.

- The different types of labour (corresponding to different wage-fixing bodies) are identified and the basis for evaluation of a unit of labour cost identified. Usually this would include the following:
  - ○ Changes in the hourly or weekly wage rate as determined by agreement of the parties to the wage-fixing body.
  - ○ Changes in employer's 'on costs' and contributions such as holidays with pay, National Insurance contributions, etc.
  - ○ Changes in the agreed standard working week.
  - ○ Changes in the average hours worked per week, as recorded in government statistics. This item and the preceding one will affect the degree to which overtime rates are paid for each unit of labour output.
  - ○ Changes in productivity. One method of obtaining a gauge of productivity is to take the total quantity of materials used by the industry, priced at 'standard rates' (i.e. constant figures) and divide by the total building labour force. If the output per operative increases over a particular time period then it is assumed that this is the result of productivity and the labour weighting should be reduced accordingly.
  - ○ Change of location. It may be necessary to assume a particular geographical position for the typical building in order that regional differentials can be ignored. The labour unit cost for each type of craftsman and labourer would be computed and then weighted according to the importance of each in the total labour force. The resultant weighted average for labour would be carried forward for weighting with other resources.

- Material rates for comparison with base year prices are selected. Although it is possible to identify a long list of cost significant materials and calculate a separate index for each, it is more usual to rely on published statistics for this type of information. The Department of Trade and Industry regularly produces indices of materials for a range of material classifications. It is then merely a question of analysing the typical building according to the published material types to obtain the weighting to be applied to each material index.

- Plant types, together with their weighting, are identified and a standard method of evaluation adopted. This may be based on an average of hire rates for the particular item or it may be calculated by considering purchase price, depreciation, maintenance, standing time, etc., to establish a standard resource unit cost which can form the basis of the index. In many published factor cost indices the change in the cost of plant is ignored.

When the weighted index for each of the three basic resource types is computed they are then brought together and weighted again, this time in accordance with the importance of each within the total building cost. Typical values for weighting are:

Labour   35–45%
Material   50–60%
Plant   5–10%

The final weighted index represents the change in cost between one point in time and another for the typical building chosen. Any subjective judgement to take into account market conditions, productivity, etc., can then be made if it is considered necessary.

Limitations and uses of factor cost indices

It is important to realise that the factor cost index is actually measuring the change in the cost of resources to a contractor for a 'typical' building. Its main shortcoming from the quantity surveyor's/cost adviser's point of view is that it takes little or no account of the tendering market. It does not directly measure the change in the price the client must pay, and although the 'market conditions' allowance attempts to rectify this situation the source may not necessarily be reliable, having usually been made on a subjective basis as suggested above. Neither does it measure the change in cost of a specific building under consideration by a quantity surveyor.

It is unlikely that the 'typical building' will have the same mix of resources as the client's project; however on many occasions the difference will not be cost significant. To try and minimise this difference the 'typical' building may be more of an economic model than a likely real-life building, having, say, a mixture of light concrete block and clay block partitions so that a change in the cost of only one of the alternatives will not unduly distort the index on the one hand nor be ignored on the other.

In spite of these drawbacks the method is very suitable for identifying trends in resource costs and relationships. It was particularly useful for evaluating cost fluctuations in contracts which allowed for reimbursement of any changes in cost to the contractor that occurred during the contract period, as was common practice at times of high inflation.

The tender-based index

A much more attractive proposition than the factor cost index from the private quantity surveyor's point of view appears to be an index which takes as its source of information the tender document itself. This should record what is happening in the market place and therefore should be much more useful in updating prices for a design budget.

In essence, the tender figure is compared with a figure produced by pricing the same tender document using standard rates at the base year prices. From the two resultant figures an index can be calculated showing the increase or decrease in cost to the client within the current tendering market.

Limitations of the tender-based index

There are drawbacks with the method, including:

- the questionable validity of BQ rates;
- the difficulty of obtaining priced BQ for jobs which are comparable except for date of tender.

If, however, a large enough sample of projects is taken then the difficulties can be dealt with by 'trampling the problem to death'. If an appreciable quantity of projects is analysed it is hoped that all other differences except current levels of cost will cancel themselves out.

An obvious difficulty is the work involved in analysing all the rates in scores of BQs. Fortunately it has been found that much time can be saved by taking only the few largest items in each work section and this results in a negligible degree of error.

Constructing a tender-based index

The procedure for preparing and using a typical tender-based index can be summarised as follows:

- Prepare a priced list of typical BQ items at the prevailing base year pricing levels (possibly with an allowance for preliminaries included). This is of course a time-consuming task and the current published forms of this index tend to use the *PSA Schedule of Rates for Building Works*, which is available from HMSO. This schedule is very comprehensive and is more than adequate for the majority of buildings.

- Take the priced document for the lowest tender received and note the format (e.g. work section, elemental).

- For each section of the format (e.g. excavator, concrete work) pick up the largest value item in the section, then the second, and so on, until a total of 25% of the section value is achieved. (It is of course the value of the item as extended in the cash column, and not its unit price, with which we are concerned.) This procedure is repeated for all sections with the exception of preliminary items, PC and provisional sums, profit and attendance on PC items, and daywork percentage additions.

- For each of the items selected find the corresponding unit price in the base schedule of prices. The difference as an addition or subtraction from the base year rate is then obtained by comparing the BQ unit price of each sample item with its equivalent base year price.

- The extended base year value is calculated for each item (i.e. weighted by its quantity). The total of all the extended items at base year prices is then compared with the total of all the extended items for each section to obtain an index.

- The index obtained for each section is then weighted according to the value of that section as a proportion of the whole tender BQ for the project being considered (less preliminaries, PC and provisional sums, contingencies, etc.) and a combined index for all sections obtained.

- The preliminaries are usually dealt with by considering them as a percentage addition to the other items in the BQ, excluding dayworks and contingencies. If the rates in the base schedule include an allowance for the preliminary items then the index figure arrived at so far is not a true one, since net BQ rates have been compared with gross base rates. So, if the 'false' index of BQ items is 130 and the value of the preliminaries in the current tender is 10% (as a percentage of the remainder of the BQ) then the final index would be:

$$130 + \left(130 \times \frac{10}{100} \right) = 143$$

- To obtain an average index for publication rather than for one particular job, the index numbers for all the tenders which have been sampled in a specified time period are averaged by taking either the geometric or arithmetic mean. The difference between the mean of the current and preceding quarters is used to gauge the movement in tender prices. The geometric mean is usually taken when relative changes in some variable quantity are averaged, because the arithmetic mean has a tendency to exaggerate the 'average' annual rate of increase.

A number of simplifications have been made in the above procedure for the purpose of clarity. For example, in practice, problems arise with items that cannot be matched (these are generally overlooked in the item selection process), and where a sample amounting to 25% of a section's total cannot be obtained. The rules governing the index being used should always be carefully consulted in such situations to discover the method to adopt.

Uses of the tender-based index

The tender-based index has several advantages over a simple factor cost index from the point of view of the client's quantity surveyor:

- It measures the change in the cost to the client of a particular project over time, taking full account of market conditions in addition to the change in cost to the contractor.
- It is relatively simple to operate once a base schedule of prices has been obtained.
- It allows comparison of the price obtained by tender for a specific project with the national or regional building price trend. This could assist in deciding whether to call for fresh tenders from another group of contractors, or not.
- It allows the relationship between the market for buildings of different function and locality to be plotted.
- It is not based on other indices and therefore any inherent inaccuracies are not compounded.

There are, however, one or two problems which have to be acknowledged:

- To overcome the problems due to the variability of price rates in BQs a large number of projects are required for each index. It has been suggested that at least 80 projects are required for a suitable sample, but this condition is seldom met. Unless access to this number of BQs is available the trend being plotted may be erratic and not typical. Very few organisations have access to this number of projects and therefore most firms cannot prepare their own index by this method.
- The index relies heavily on the base year schedule which will have to be regularly revised to take into account new products and new methods of measurement. This is a time-consuming and costly task.
- Because of a lack of suitable projects at any one point in time the average index may have to rely on an unbalanced sample containing more jobs of one particular function and location than is considered desirable. This may lead to errors in the trend plotted.

However, despite these problems there is no doubt that the tender-based index represents a major breakthrough in the measurement of trends in market prices to the client.

Published forms

There are a large number of published indices which are available to the construction industry and the main sources are listed below, classified under the two major index types.

Factor cost indices

Of these perhaps the following are the most familiar.

NEDO price-adjustment formulae

The one-time National Economic Development Organisation (NEDO) produced a series of indices for building trades based upon the resource costs of each trade. These are now maintained by the Construction Directorate of the DETR and are published by HMSO. They are still popularly known as the NEDO formulae, although the correct name is the Price Adjustment Formulae for the construction industry.

The indices are based on market prices, nationally agreed wages, etc., and therefore do not take into account the problems of a contractor on a particular site.

Although primarily provided for the assessment of price fluctuations on contracts where there is a fluctuations clause they can be adapted for use in a composite building cost index, as in the BCIS index dealt with next.

BCIS building cost indices

These measure the change in basic input costs – labour, materials and plant – and are based on cost models of average buildings. There are nine indices, as follows:

- BCIS General Building Cost (exc M&E) Index
- BCIS General Building Cost Index
- BCIS Steel Framed Construction Cost Index
- BCIS Concrete Framed Construction Cost Index
- BCIS Brick Construction Cost Index
- BCIS Mechanical and Electrical Engineering Cost Index
- BCIS Basic Labour Cost Index
- BCIS Basic Materials Cost Index
- BCIS Basic Plant Cost Index

The model for each BCIS Index is based on the Price Adjustment Formulae for Construction Categories (Series 2) published by the DETR.

The BCIS indices are calculated by applying the work category indices to the models. Each work category index is a compound of resources of:

- labour,
- materials and
- plant

and changes in the cost of each of these components is reflected in the work category indices.

Generally, the labour indices are based on the national labour agreements and the materials indices are based on the Producer Price Indices prepared by the Office for National Statistics.

The models were prepared by analysing 54 new work BQs into the building work and specialist engineering work categories. The BCIS General Building Cost Index is a weighted combination of:

- the Steel Framed Construction Cost Index (34%);
- the Concrete Framed Construction Cost Index (8%);
- the Brick Construction Cost Index (58%).

These proportions were derived from a survey of 837 jobs from the BCIS *Online* database with tender dates between 1989 and 1993, which showed that:

- 287 jobs were of steel framed construction;
- 64 were concrete framed;
- 481 were traditional brick construction;
- 4 were timber framed;
- 1 was unspecified.

The BCIS General Building Cost (excluding M&E) Index is a similar index to the BCIS General Building Cost Index, but excludes the following installations:

- electrical;
- heating, ventilating and air conditioning;
- sprinkler;
- lift;
- catering equipment;
- other specialised M&E items such as fire alarms.

The cost indices for steel framed construction, concrete framed construction and brick construction are all based on cost models containing buildings with each particular form of construction only. They include all categories of work.

The weightings obtained for the General Building Cost Index and the Steel Framed, Concrete Framed and Brick Construction Cost Indices are shown in Table 7. The weightings are at June 1976 levels to correspond to the date at which the weightings for the DETR indices were calculated.

In principle, to construct the index for any point in time, it is only necessary to multiply the DETR index for each work category by the weighting, sum the extended totals and divide by 100. However, in practice the calculation is complicated by the fact that the DETR work category indices do not all have the same base date. This is further complicated as the BCIS series of Building Cost Indices currently have a base date of 1985 = 100.

The calculation of an index is therefore ideally suited to calculation by computer, as it would be almost impossible to do manually.

'Building' housing cost index

This index is based on fixed weightings of specified labour, materials and overheads found in typical house building and is published in *Building* magazine.

Table 7   Weightings for BCIS Cost Indices.

Work categories	% Weightings at June 1976 levels			
	Steel framed cost index	Concrete framed cost index	Brick construction cost index	General building cost index
2/1 Demolitions	0.69	0.75	0.56	0.61
2/2 Site preparation, excavation & disposal	4.26	2.21	4.91	4.48
2/3 Hardcore & imported filling	2.22	0.82	2.49	2.27
2/4 General piling	0.41	2.74	0.24	0.50
2/5 Steel sheet piling	0.00	0.02	0.00	0.00
2/6 Concrete	5.66	5.79	4.70	5.11
2/7 Reinforcement	1.46	5.44	1.09	1.56
2/8 Structural precast & prestressed concrete units	1.13	0.67	1.74	1.44
2/9 Non-structural precast concrete components	1.21	0.62	1.81	1.51
2/10 Formwork	1.42	5.70	1.10	1.59
2/11 Brickwork & blockwork	6.53	3.83	12.56	9.81
2/12 Natural stone	1.00	1.46	0.59	0.80
2/13 Asphalt work	0.03	0.46	0.10	0.11
2/14 Slate & tile roofing	0.49	0.45	2.58	1.71
2/15 Asbestos-cement sheet roofing & cladding	0.00	0.01	0.02	0.01
2/16 Plastic coated steel sheet roofing & cladding	5.80	1.68	0.50	2.39
2/17 Aluminium sheet roofing & cladding	1.39	0.77	0.26	0.68
2/18 Built-up felt roofing	0.12	0.26	0.31	0.24
2/19 Built-up felt roofing on metal decking	0.00	0.08	0.00	0.01
2/20 Carpentry, manufactured boards & softwood flooring	2.22	1.71	7.53	5.27
2/21 Hardwood flooring	0.12	0.06	0.16	0.13
2/22 Tile & sheet flrg (vinyl, thermoplstc, lino & other synth mats)	1.96	1.54	1.69	1.77
2/23 Jointless flooring (epoxy resin type)	0.22	0.08	1.04	0.68
2/24 Softwood joinery	2.21	2.52	5.35	4.05
2/25 Hardwood joinery	0.84	0.82	1.63	1.31
2/26 Ironmongery	0.92	1.07	1.77	1.43
2/27 Steelwork	0.02	0.49	1.51	0.93
2/28 Steel windows & doors	0.67	0.71	0.14	0.37
2/29 Aluminium windows & doors	2.52	6.09	1.16	2.02
2/30 Miscellaneous metalwork	3.75	4.93	3.35	3.61
2/31 Cast iron pipes & fittings	0.33	0.47	0.13	0.23

Contd

Table 7 Contd.

Work categories	% Weightings at June 1976 levels			
	Steel framed cost index	Concrete framed cost index	Brick construction cost index	General building cost index
2/32 Plastic pipes & fittings	0.66	0.12	0.56	0.55
2/33 Copper tubes, fittings & cylinders	0.53	1.07	0.48	0.55
2/34 Mild steel pipes, fittings & tanks	0.05	0.03	0.22	0.15
2/35 Boilers, pumps & radiators	0.05	0.04	0.73	0.44
2/36 Sanitary fittings	0.48	0.62	1.47	1.06
2/37 Insulation	0.51	0.74	1.07	0.85
2/38 Plastering (all types) to walls & ceilings	0.63	0.67	2.31	1.60
2/39 Beds & screeds (all types) to floors, roofs & pavings	0.34	0.45	1.03	0.76
2/40 Dry partitions & linings	1.44	1.29	1.33	1.36
2/41 Tiling & terrazzo work	1.26	0.59	0.94	1.03
2/42 Suspended ceilings (dry construction)	1.10	1.96	0.87	1.03
2/43 Glass, mirrors & patent glazing	3.49	2.54	1.21	2.09
2/44 Decorations	1.03	1.00	2.45	1.85
2/45 Drainage pipework (other than cast iron)	1.27	0.36	1.41	1.28
2/46 Fencing, gates & screens	0.41	0.12	1.02	0.74
2/47 Bituminous surfacing to roads & paths	1.01	0.23	1.39	1.17
2/48 Soft landscaping	1.31	0.17	1.36	1.25
2/49 Leadwork	0.15	0.14	0.36	0.27
Electrical installations	10.42	10.07	8.89	9.50
Heating, ventilating & air conditioning installations	10.23	18.19	7.51	9.29
Sprinkler installations	0.00	0.11	0.00	0.01
Lift installations	1.05	3.54	0.84	1.11
Catering equipment installations	0.64	0.40	0.07	0.29
Structural steelwork	12.34	1.30	1.46	5.14
	100.00	100.00	100.00	100.00

There are a number of other cost factor indices which appear in the technical press and builders' price books. The derivation of the indices is not always known and therefore they have not been included here, and should be used with caution.

Tender-based indices

As tender-based indices are a comparatively recent innovation there is not the long history of results that could be found with factor cost indices. However, the number of institutions and practices producing an index of this type is steadily growing.

DETR (Construction Directorate) index of building tender prices

The construction of the index is similar to that already described in this chapter, where a sample of items is taken for each work section up to 25% of the value of the particular trade.

This index was the original tender price index and has formed the basis for the others which have followed.

BCIS tender-based index

The method of construction is similar to the DETR form just described. Two indices are prepared, one relating to fluctuating contracts and the other to fixed price. This goes some way towards overcoming the problem of the different allowances made in tender figures for future variations in price.

The sample of jobs from which the index is derived is not restricted to government projects, and therefore it can be expected to cover a wider range of building types than the preceding index.

For a small fee the BCIS will undertake the preparation of an index for a priced BQ from any of its members in order that a tender can be compared with the national average.

Davis Langdon and Everest tender price index

Again, the method of construction is similar to the DETR index, the difference being that this index:

- is based upon schemes in the London area only;
- includes non-government jobs;
- is probably based upon a smaller sample.

The index is published quarterly in *Building* Magazine.

Other indices are available which are in general use but do not strictly fall under the previous two headings. One example is the Nationwide Building Society's Index of House Prices, available on the Internet. This is based on mortgage information and measures the change in cost to purchasers of different house types in different regions of the UK. It should be used for guidance purposes only as house prices vary widely depending on such factors as size, location, specification and fashion; it is, of course, substantially affected by variations (usually upward) in land values which may be considerably greater than building cost fluctuations.

Another example is the DETR price index of local authority house building, where a number of priced items are extracted from BQs for traditional one- and two-storey houses in each quarter, and the index prepared from the average change in price of these items only. The items are selected on the basis of major cost significance and factors are applied to each item according to base year weightings, thus following the approach of a Laspeyres index.

To obtain a national price index it is necessary to employ fixed regional weights, not only because the amount of building varies greatly from one region to another, but because price levels, and the rates of change in prices, vary too.

Comparison of index performance

The degree to which an index performs or moves depends on two factors:

- the market factors influencing the components represented within the index;
- the weightings applied to the components within the index.

This is true whether we are considering a tender-based or a factor cost index. In the latter case it is comparatively easy to see that if you measure an increase of 20% in the price of one material then, if all other factors remain unchanged, the index will go up by an amount equivalent to the weighting of that material in the total index, multiplied by 20%.

With the tender-based index this process is obscured by the contractor's pricing method, and the fact that material cost is not shown as a separate item. However, the increase in the cost of this material will be included in the unit rates by the contractor and will be subsequently weighted by the quantity of the items containing that material. It will also be weighted a second time when the work sections themselves are weighted.

The tender-based index will, in addition, include for any extra charge in the rates for increased overheads, profit, or above normal resource costs due to the current market situation. It is these features that produce the distinctive performance of a tender-based index compared with the factor cost type.

By way of comparison let us look at the performance of the BCIS General Construction Cost Index (factor cost type) with the BCIS Tender Price Index (tender-based) for the years 1978 to 1997. The movement of each index is shown in Fig 14.1 superimposed upon the industry activity indicators of contractors' new orders and contractors' output of new work expressed at constant 1990 prices. It will be noted how steady the rise in the Construction Cost Index was during the whole of the period, the graph showing what is very nearly a straight line. However, the tender-based index characteristically reflects a more erratic trend in the cost to the client.

Towards the end of the 1970s the then Labour government was maintaining a substantial government capital programme, but with the return of the Conservatives in 1979 and their emphasis on monetary controls to defeat inflation, government expenditure on building was cut yet again. The impact can be seen in the stabilisation and even fall of price to the client in 1980–82, whereas resource costs continued to rise.

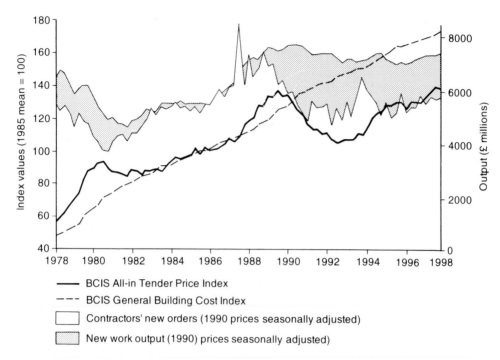

Fig. 14.1   Relationship of cost indices to trends in new orders and output of contractors' work. (Reproduced with kind permission of the Building Cost Information Service.)

However, a major programme of privately funded building through the 1980s saw the two indices in very close agreement for a number of years, until with a major increase in new orders in the late 1980s tender prices began to climb again. This was a short-lived boom – the crash that followed it saw a real fall in tender prices while actual costs to builders were increasing, accompanied of course by substantial redundancies and bankruptcies.

However, contractors cannot sustain their operations for ever by absorbing the increase in resource costs from their reserves, and at the time of publication there are signs that tender prices are rising again.

The reader will notice a strong correlation between the industry's output (especially the start of new projects) and the tender-based index, suggesting that present and perceived future work load has a very strong influence on the price to the client.

In any comparison of index performance the important fact that must be realised is that each index is measuring a different movement of cost or price. Even where the form of the index is the same, the difference in the weightings attributed to each component by individual indices will result in a different value for each at the end of all the calculations. In using an index it is essential that users are quite clear which trend in cost they wish to measure, and then pick the form and weighted structure that best suit their objective.

Problems in constructing and using cost indices

It has already been stated that there are a number of problems associated with the construction of cost indices. Such problems can also create difficulties in the use of the index and these, together with some more general points, are summarised below:

- The index will usually be measuring a trend for a typical or model building type, and may not necessarily be measuring the change over time for the particular project that is being developed.

- Where the base date is several years old then the question should be asked whether the index is being based on outdated criteria. This particularly applies to a factor cost index using base year weightings. It may be that the balance of resources has changed considerably in the intervening period and that this is not represented in the index. Over quite a short period the 'mix' of resources on a typical building may alter as a result of the very changes in cost that the index is trying to record; for instance a rise in metal prices may cause a substitution of plastic for lead and copper in plumbing and roofing work.

  In the longer term, changes in technology may not only make the base 'mix' untypical but may also distort its prices – as new techniques and materials come into greater use their price tends to decrease proportionately, whilst the cost of obsolescent technology tends to rise faster than the general rate of increase. However, every index must have some history in order that trends can be identified.

- Regular publication of the index at monthly or quarterly intervals is required. Otherwise the user is put in the position of having to forecast the immediate past (i.e. the time between the last publication and the present) as well as the future.

- The basis of construction of the index must be known in order for it to be used intelligently. The choice of a tender-based or factor cost index will be dependent on what the user wishes to measure, but in addition the weightings of any index are important in judging whether the results can be applied to the project under consideration.

- Short-term changes in an index must be treated with caution since the inherent errors in the system of compiling the index may well be the equivalent of several points on the scale.

- In the tender-based indices, a good sample of BQs must be used if bias due to regional variation and building function is not to distort the results.

- When plotting a trend in an index it is sometimes advisable to use a logarithmic scale for the index against a natural scale for time. The advantage of this method is that if costs are rising at a regular percentage every year then the index values will be shown as a straight line. Any deviation above or below this line will show immediately as a change in this regular pattern and

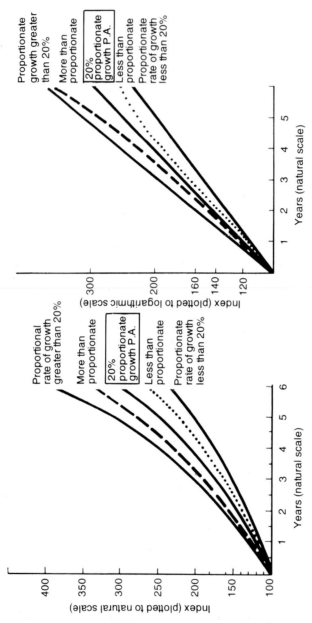

Fig. 14.2 Showing typical cost increases plotted against natural (LH) and logarithmic (RH) vertical scales. A proportionate rate of growth is one which increases annually by a constant percentage of its immediately previous value (e.g. an increase of 20% on an index value of 100 gives 120. A subsequent increase of 20% would give 144, not 140). Note the characteristic shape – straight, concave, convex – of the index graphs in the second diagram, making them much easier to identify at a glance.

will assist in detecting a new trend. Figure 14.2 shows the principles involved using a 20% proportionate increase as an example. If a natural scale is used for the index then an exponential curve will be produced by the 20% increase and comparison is more difficult.

Cost planners obviously have a duty to keep themselves aware of past, current and future economic trends, and should use this awareness in coming to a conclusion on any extrapolated index value which they wish to use in estimating for future projects. Of course no one has yet proved that they can accurately forecast the future, and this particularly applies to future index numbers. Anybody who could actually do this would be in the City earning a multi-million pound salary, not cost planning the average building project.

Providing reasonable care has been exercised, cost planners should not be blamed if their forecasts turn out to be wrong due to changed economic or political factors that were unforeseen at the time of making the estimate. For this reason many cost planners tend to use a reputable index such as the BCIS 'All-in' Tender Price Index, on which any responsibility for 'getting it wrong' can be placed. This might be more difficult if an in-house or less well known index were used.

In recent years it has become common practice and far more sensible to give a range of anticipated future costs based on the possibility of certain events occurring. In any case, it is most important that everything should be made explicit when the original estimate is given. The details should include:

- the index which has been used;
- the point or range that has been chosen together with the reasons for this choice;
- the time scale for the project which has been assumed when making the forecast.

Which type of index to use

By their very existence it is obvious that all cost indices have a role to play in assisting the cost adviser to determine the market forces affecting a project. The choice of index will, however, be determined by the use to which the index is put:

- For short-term forecasting, allowances for fluctuations, etc., the *factor cost index* can be used.
- For general estimating and cost planning purposes a *tender-based index* will more reliably measure the change in the cost to the client. It also has the advantage of being able to measure the performance of a tender, region or building type against the national average.
- For very long-term forecasting and budgeting the use of a *unit price index* based on cost per square metre or unit accommodation may be more applicable.

Summing-up

The object of a cost index is to measure changes in the cost of an item or group of items from one point in time to another. The inflation index produced by the government every month is a well known example.

In compiling an index a base date is chosen and the cost at that date is usually given the value of 100, all future increases or decreases being related to this figure.

Published cost indices are available for:

- building in general;
- various types of building;
- various components of building cost.

Building cost indices may index either:

- construction costs based on the cost of resources or
- tenders.

These differ because tenders are more affected by market conditions.

The use of a reputable published index gives the cost planner some protection if unforeseen events occur.

Chapter 15
Cost Planning the Brief

The brief

After the initial budget has been prepared and accepted (Chapters 7 and 8) the cost planning process can begin in earnest, with the brief being agreed and outline drawings being prepared.

An iterative process

It might be thought that the title of this chapter is misleading – since the cost plan itself is not going to be prepared at this stage a better title would surely be 'Costing the Brief'?

This might be so if that was all that the cost planner was going to do – passively estimate the cost of carrying out the client's ideas.

But this will not be good enough. We have seen that this is the stage where the big cost decisions are made, consciously or unconsciously, and design work should not start until the brief itself is right. There is little point in spending large sums of money meticulously – and successfully – cost planning and controlling a design which is not in fact a very good answer to the client's needs.

In the case of profit projects it is often possible to charge higher rents or get a higher price for leases if a better standard of amenity and finish is provided, so that the budget is flexible to a limited extent. In such circumstances the cost disadvantages of larger and more luxurious public spaces, more and faster lifts, better fittings, and so on, will have to be evaluated in relation to income. The acceptable final cost may be quite different to the original rough estimate due to such decisions.

Clients for private residences often become more ambitious as the drawings progress, and again provided that they are kept informed of the consequences of their decisions they may prefer to have the house they now want rather than the cost they first thought of.

It is obviously preferable that these matters should be decided at brief stage rather than during design development or, even worse, during construction.

Many authorities consider that the traditional practices of the British building industry make it too easy for the client to make changes during the construction stage and that this is partly, or even largely, to blame for the comparatively high cost of building in the UK compared to some other countries. The cost planner therefore must be helping to develop the brief and examine its cost implications critically (unless specific instructions to the contrary have been given – but clients rarely do that these days!).

Nevertheless, the first task is indeed to cost the brief as presented. With sketchy information and without any sort of design it will be impossible to use any kind of resource-based cost model, so even where it is proposed that the final cost plan will be based on resources there is little alternative to a product-based estimate at this first stage.

Preliminary estimate based on floor area

This will be a very approximate estimate; it could hardly be otherwise in view of the lack of information at the cost planner's disposal. There are exceptions to this, for instance:

- Where the project is one of a series (such as a chain of service stations or superstores) and information has been accumulated on a number of almost identical buildings.
- Where the first estimate has to be calculated according to a government formula and the building will be tailored to fit the estimate.

In the preparation of the preliminary estimate based on floor area, previous experience of the designer is most valuable as there are some whose designs always come out above average costs just as there are those who can be relied upon to achieve an economical design.

At an early stage the cost planner is quite likely to be told 'This will be an economy job' and only a knowledge of the people or practices concerned will indicate how far this statement can be relied on in framing the estimate.

The cost planner should never allow the amount of the estimate to be influenced by opinions expressed by interested parties, or worse still by the figure that the client would like the job to cost, unless these also coincide with the cost planner's own judgement.

When an estimate based on early designs has worked out at £3,500,000 it is all too easy to remember that the budget is £2,250,000. The danger then is that the cost planner may assume that the estimate is too high and reduce it a little in order to make the difference less breath-taking.

Cost planners may also find themselves in front of a committee or a board, one of whose members 'knows a job just like this one which was built for much less than the cost planner's figure'.

Remember that it is the person who has prepared and possibly signed the estimate who will be held responsible for it, not the designer, client or the assertive committee member. This is as it should be. Cost planners are paid for their skilled evaluation of real probabilities, not for telling people what they want to hear.

The initial estimate, like all others, must make quite clear what is and is not included. It is quite good practice to use a form for preparing the estimate, with a definite section to show inclusions and exclusions.

A computer spreadsheet program should similarly present a menu of possible inclusions and exclusions to jog the cost planner's memory, and to print them out when preparing the estimate for the client. Such items as:

- site preparation, such as demolition or dealing with contaminated land,
- archaeology,
- complete internal decoration,
- joinery fittings,
- venetian blinds and curtaining,
- furnishings,
- office partitioning,
- lighting fittings,
- carpeting in offices and flats,
- site works,
- specialist plant and services,
- shopfitting and
- security systems

should be specifically dealt with, not just forgotten about or covered by some vague overall clause. In fact it is even more important to define these things properly in a preliminary estimate than it is later on, when the detail of the estimate itself may be sufficient to answer many of the questions.

Make sure that professional fees, VAT and project manager's and construction manager's fees (where appropriate) are clearly excluded or included – clients quite rightly think that the estimate shows their total commitment unless they are told otherwise. It is especially important to make the position about VAT clear – it is the usual current practice for cost planners and estimators to ignore this, but the client must not be left in any doubt.

If fees are shown it is important to get them right. Until the mid 1980s this was comparatively easy, as everybody charged according to the fee scales of their various professional bodies. Today, when fees are subject to negotiation or competition, this is more difficult, but unless something has already been agreed to the contrary it will probably be best to assume fees according to scale at this preliminary stage. Reductions can always be made later, if necessary.

Professional fees should be estimated to project completion, not just design completion, and should include the various specialists as well as expenses and document reproduction. Despite the supposed advent of the 'paperless office' this last can be a considerable sum.

The estimate must also be specific about inflation. If it is based upon current building costs it should say so, and also make clear that it is subject to revision in the event of increases or decreases in those costs. There have only been one or two periods when building costs have been fairly static in the last 40 years. The longest such period has been in very recent times so that there is a tendency not to get too worried about this, particularly among students and trainees who have never experienced major escalation in building costs. However, even if things seem to be set fair at the time when the estimate is being prepared it is still wise to mention that it is based on current costs and may need adjustment for inflation.

If, on the other hand, the estimate is based upon anticipated cost levels in the future the basis of calculation should be set out in the estimate so that revision can be made if the prediction proves to be false. But in both cases there are advantages in tying the estimate to a well known and reputable cost index such as the BCIS index, so that the cost planner's personal judgement cannot be blamed if things suddenly change.

As has already been pointed out, the person who could accurately forecast economic conditions 2 or more years ahead would be in demand in other and more lucrative fields than construction cost planning!

A final problem relates to whether the tender or the final cost is being forecast. Clearly the latter is the preferable alternative, since it is this sum of money which the client is going to have to find in the end, and in any case there is little alternative in the case of projects which are to be built under a contractual arrangement that does not involve a lump sum tender as such.

However, in the public sector in particular the custom is often to estimate the tender amount, and it is important in both public and private sectors to be specific about which alternative has been used. At times of high inflation there can be a considerable difference between the two figures, especially if the cost planner makes some allowance for the claims which often arise during the course of the work and which will be reflected in the final cost but not in the tender amount.

As the preliminary estimate cannot be related to the subsequent cost plan (except in total) it is not worth preparing it in any greater detail than is necessary for its primary purpose, that is to give an approximate estimate of cost. Also, although cost planning is very much more than the preparation of an estimate, it gets off to a bad start if the first estimate is badly wrong. This stage is therefore extremely important.

Finding the right rate at which to price floor area is not easy, and the most reliable starting point is a recent similar job from office experience. Failing this, or as a check on it, a wide selection of costs from a published source such as the BCIS system should be used, resisting the temptation to choose only from the lower part of the cost range. The BCIS *Online* computer system is a particularly good source, not least because of the way in which it facilitates the adjustments which are about to be described.

To adapt the floor area rate from one project to another when shape is not known requires consideration of seven factors, as follows:

- market conditions;
- size, number of storeys, etc.;
- specification level;
- inclusions and exclusions;
- services;
- site and foundation conditions;
- other factors.

A most important point is to check what is actually meant by floor area. This might be:

- The total area at each floor level measured over the external walls including all staircases, lift wells and similar voids, and including all internal and external walls.
- The gross internal floor area (GIFA) measured according to SFCA/BCIS rules. This is the same as the previous example, but excludes the thickness of external walls.
- Something rather less than the last, perhaps excluding voids, circulation spaces or plant rooms.
- The net commercially lettable floor area (often still given in square feet to complicate the issue).

Obviously for the same building these would each yield quite different cost factors, and it is important to check that the same option is used for all the data source examples and the new estimate, or that conversion factors are duly applied.

Further points for care include factors which may not be properly reflected in traditional floor area formulae, and which may occur either in the project being estimated or in the cost data examples. The most troublesome of these is the atrium, which was not a common design feature when these floor area formulae were devised, and which if large cannot be treated as an ordinary void if comparison is being made with non-atrium buildings.

A somewhat similar problem arises with buildings which have an unenclosed ground floor used as part of an open car parking area.

An example of a preliminary estimate

We can now proceed to work through an example in which the above factors are taken into account where applicable.

The proposed project is a new 2,400 m^2 six-storey office block in a provincial city, with shops on the ground floor and offices on the other five, with a budget of £2,000,000. The previous project on which the estimate will be based is a project of similar quality in outer London, but of four storeys only and also with shops on the ground floor. The total floor area was 1,235 m^2 and the tender price 12 months previously was £680/m^2.

The differences between the two schemes in this textbook example are rather more than one would ideally like to see when choosing a suitable cost plan to adjust, but will enable all the issues to be considered and demonstrated.

Market conditions

This adjustment deals with the changes in building prices and in tendering conditions between the time when the previous job was priced (or built) and the anticipated date of tender (or of building) for the new project. Two quite different issues are being considered together here:

- official changes in the cost of resources;
- changes in the market itself, taking into account the capability of the industry and the amount of work available to it, in the regions and at the periods in question.

In doing this it is most important to compare like with like. It may be, for instance, that the project being used as a basis for comparison was tendered for conventionally, so that the prices are those ruling before construction started, whereas the new job might be built using construction management where prices ruling at the time of carrying out the works are what matters.

One of the published cost indices previously described would probably be used, but where the estimates which are being compared are not too far apart in time (up to a year or so) it is quite possible to make the adjustment for resource costs by simply adding a percentage based on known labour and material fluctuations. It may indeed be necessary to do this for short-term estimates because of the time lag in publishing historical indices.

For this purpose the proportions of labour, material and plant content given in Chapter 14 could be used, with a heavier bias towards materials if the contract is for an expensive building with good finishes and high-class materials. In making this calculation the figures should be taken as fractions of the gross contract amount without deducting overheads and profit, since these should rise more or less in proportion to increases in prime cost.

Where 'do-it-yourself' indexing of this type is being used, or where a published index is being used which only reflects changes in the prices of resources, it is most important to remember to make the adjustment for changes in market conditions as well. This can be done intuitively if both projects are local, but if one or both of them is in an area which is unknown to the cost planner it may be necessary to make discreet enquiries.

Local changes in the market often tend to be relatively short-term (because an overheated local construction sector just has to cool down somehow), and thus are not always easy to pick up from published building cost indices. Local enquiry may be worthwhile in a case such as the present one if the new project is in an unfamiliar region.

In this particular example allowance would have to be made for the difference between outer London and the provincial city, remembering whether or not differences in resource costs (e.g. labour rates) have already been adjusted.

Let us assume an overall upward adjustment of 10% for inflation and region on this contract. This can either be shown at the beginning of the estimate, thus:

£680/m^2 plus 10% = £748/m^2

or added on at the end.

The first method is preferable, as all the adjustments which follow can then be done at current prices, which is usually easier.

Size, number of storeys, etc.

At this stage no actual drawings or dimensions are likely to be available, so that adjustments have to be done by proportioning, that is the use of ratios in comparing the project used as an example and the new project.

In the early days of cost planning it was fashionable to use proportioning methods extensively, to show that the new approach was different to old-fashioned approximate estimating. But this point no longer has to be demonstrated, and proportioning has two disadvantages:

- It is not the quantity surveyor's usual method of working and thus mistakes are more easily made.
- Sooner or later approximate quantities have got to be used, and these are difficult to reconcile with proportioning.

It is therefore recommended that proportioning only be used when there is no alternative.

The new provincial block will have a larger plan area than the London example; this will tend to reduce the proportion of external wall area to floor area (assuming a similar plan shape) and hence should reduce the cost per m^2 of floor area a little.

However, the new block will have six storeys instead of four, and this will have the effect of increasing the cost as it is a general rule that building higher costs money. But the difference between four and six storeys is not critical, and the likelihood is that this and the previous factor will cancel each other out.

The less that is known about the new building, the less point in adjusting for hypothetical trifles. We are working with a broad brush at this stage, trying to get a 'ball-park' figure as a starting point for the cost planning process. However, a more important point to consider is that on the outer London block one-quarter of the floor area was set aside for shops whereas on the six-storey building the figure will be one-sixth.

As shop areas are normally left in an unfinished state for the shopkeepers to fit out according to their needs and tastes ('shell and core') this part of the building will be cheaper than the rest. By reducing the proportion of the building allocated to shops we shall be increasing the overall price/m^2.

Assuming that the saving in floor, ceiling and wall finishes, and services between offices and shops will be £72/m^2 of total floor area:

Outer London block

Shops (25%)

£748/m^2 minus 3/4 of £72 = £694/m^2

	309 m^2 at £694/m^2 =	£214,446

£748/m^2 plus 1/4 of £72   = £766/m^2

	926 m^2 at £766/m^2 =	£709,316
		£923,762

Check – total 1,235 m^2 at £748/m^2 =   £923,780

Provincial block

Shops	16.67%	400 m^2	at £694/m^2	=	£277,600
Remainder	83.33%	2000 m^2	at £766/m^2	=	£1,532,000
					£1,809,600

£1,809,600 divided by 2,400 m^2 = £754/m^2

Similar adjustments may be required wherever the buildings are divided into areas with different cost profiles and where the proportions are different on the two buildings. Car parking areas within office buildings frequently lead to adjustments of this sort.

Specification level

So far the changes have been dealt with by adjusting the elemental rate, as there has not really been any alternative and this was in any case the preferred approach in 'classical' elemental cost planning. But from now on it is possible to think in terms of lump sum changes and adjust the elemental rate at the end, and many cost planners prefer to do this. We will show both methods.

Whichever method is used, it is important to remember whether or not the building cost has already been adjusted for inflation. If it has been (as is recommended), then any specification adjustments must be done at prices ruling at the date of the new scheme, and not those ruling when the previous or example project was carried out. Alternatively, of course, all specification adjustments could be made at the earlier pricing level and the inflation adjustment for the total scheme done at the end.

It is assumed that the specifications of the two buildings will be of the same general standard, but:

- The designer has decided that the cheap floor covering used in the offices in outer London was a mistake, and wants to allow an extra £12/m^2 for this.
- The client wants natural stone dressings in lieu of the cast stone used on the previous project.

The adjustment for the flooring is easy. Allow an extra £12/m^2 over the area of the offices, say two-thirds of the total floor area.

1,600 m^2 at £12/m^2 = £19,200.

This is equal to £8/m^2 of gross floor area when divided by 2,400.

Calculations like this can be done by simple proportion if preferred (1600/2400 of £12.00 is £8.00), but it is quite easy to make a mistake doing this – inverting the ratio for example! – and the longer calculation is more easily checked.

The stone will be a little more difficult to allow for (in the absence of any drawings, remember). The area of external walling in a medium-rise office building of this sort should be about 10% more than the floor area (again, check this with several analyses if possible). Not much of the wall area will actually be stone – most of it will probably be windows. We will assume that 25% of the area is stone and that natural stone facing will cost £130/m^2 more than cast stone. So the extra allowance for natural stone is:

(110/100) × 2,400 x 1/4 × £130.00 = £85,800

which is £35.75/m^2 of gross floor area.

Note that a number of assumptions have been made about the London example which could have been avoided if a full SFCA/BCIS type of cost analysis had been available, even though it is not intended to produce this first estimate for the new building in elemental form.

Inclusions and exclusions

The client has found that office tenants prefer to do their own partitioning and decoration, whereas on the outer London project these were provided by the developer. This adjustment will most easily be carried out by extracting the relevant lump-sum figures from the BQ for the London project, if these are available.

	£
PC sum for office partitioning	26,250
Add for profit and attendance	1,313
Decoration in offices	4,300
	31,863

As it comes from the original scheme this figure must be adjusted for inflation, also for five-sixths of the building being offices in lieu of three-quarters.

£31,863 plus 10% inflation allowance (from above)= £35,049
Division by floor area of offices on London building (926 m^2) gives a cost of £37.85/m^2 of office floor area
The saving on the provincial block by omitting these items is therefore £37.85 × 2,000 m^2 = £75,700 in total
£75,700 divided by 2,400 m^2 gross floor area = £31.54/m^2

By the (alternative) proportion method the calculation is:

£31,863 divided by 1,235 m^2 = £25.80
£25.80 plus 10% inflation allowance (from above) = £28.38/m^2

The omission of partitions in the new building per m^2 of gross floor area is therefore £28.38 × 4/3 × 5/6 = £31.53/m^2.

This all assumed that the cost of partitions was available from the BQ or cost analysis for the outer London block. If this was not the case then a telephone call to a firm of partition installers might be the only way of finding out a cost/m^2. There is no need to tell them that you want to know this so that the partitions can be omitted! This would be a current cost of course, so does not need to be adjusted for inflation or market changes.

Services

Services form a major part of the cost of any modern building. On a project of any size there will be consulting engineers for the heating, air-conditioning, plumbing and electrical work, and any specialist service installations. The terms of their appointment should include for providing an estimate of the cost of the services for which they are going to be responsible.

The building cost estimate would be adjusted as follows:

Provincial office block
Estimated figures received from consultants:

	£
Air-conditioning and space heating	165,000
Plumbing	52,000
Electrical installation	108,000
Lifts	118,000
	443,000
Builder's work, attendance, profit 7.5%	33,225
	476,225

£476,225 divided by 2,400 m^2 = £198.42/m^2

Outer London block   £
Comparable figures extracted from

cost analysis or BQ	186,000
Add 10% for inflation as main estimate	18,600
	204,600

£204,600 divided by 1,235 m^2 = £165.66/m^2
Increase in cost of services compared to
  outer London block   £198.42/m^2 – £165.66/m^2 = £32.76/m^2

In view of the early stage at which this estimate is being prepared the engineering consultants may not feel able to give an estimate for services, but it is important that they should do so. This will avoid the situation which might arise if the cost planner were to assess the cost independently, and the amount required by the engineering consultants when they prepared their detailed scheme proved to be very different. The resulting arguments would do little to improve either the harmony within the professional team or the client's confidence in the team as a whole.

Nevertheless, it is quite possible that the consultants, having been asked to establish a preliminary cost target before they have fully established what is required, will play safe and allow themselves a sum of money sufficient to cover almost any eventuality. (It is, after all, in the nature of engineering design to allow for the likely 'worst case'.)

The situation where engineering services appear to be taking rather too large a slice of the whole cake is common enough to be met by most cost planners from time to time, and how it is dealt with depends very much on the cost planner's terms of appointment. If these include a measure of executive authority then the cost planner might well call a meeting to thrash matters out; otherwise it may be only a matter of pointing out the problem to the architect, project manager or whoever is directing the project.

The cost planner should bear in mind, however, that prices per square metre for services are extremely difficult to estimate (there is still a lack of cost data in this important area), and should try to cooperate with the various consultants rather than raise difficulties. It may, for instance, be very reasonable for the engineer to insist on client requirements being defined with more precision before even a rough estimate can be given.

If there are no consultants appointed, or if at this stage they are not prepared to commit themselves, the cost planner would be better not making any adjustment to the estimate for the services, unless there are glaring differences between the outline requirements for the two buildings.

It is, however, becoming increasingly common for professional cost planners, particularly in the largest firms, to be closely involved in the cost planning of the engineering services which, after all, represent a considerable (and increasing) proportion of the budget for which they are responsible.

Special services, especially transportation services such as lifts, should always be adjusted, as their cost bears little relation to the area of the building and they are much more easily priced on a 'per unit' or 'per installation' basis. The adjustment should be done as a comparison of lump sums, as shown above.

Site and foundation conditions

These are best adjusted on the basis of the so-called 'footprint area' or site area of the building. On most buildings this should be more or less the same as the roof area.

A major factor today is the possibility of having to take measures to deal with a contaminated site. The cost planner must make enquiries of the client, the structural engineer or the local authority to find out if there are any such requirements. This is assumed not to be the case with both the sites under consideration in our example, which are in developed urban areas, but the new provincial project will require piled foundations which were able to be avoided in outer London. Either the cost of piling per m^2 of footprint area can be obtained from a suitable analysis, or a certain number of piles can be allowed for, as follows:

> Say 90 No piles average 4 m long at £75 m = £40,500
> £40,500 divided by 2,400 m^2 = £16.87/m^2

Note that the complex of pile caps and ground beams in connection with the piles will cost little less than conventional foundations, so there is no saving to be set against the cost of the piles.

Other factors

The client is anxious to have the building erected as quickly as possible and the contract period will consequently be very short. This will tend to increase the cost, both because the site labour force will need to be larger than is economical and because the number of local contractors and sub-contractors who can be relied upon to work to such a tight timetable is restricted and competition will therefore be limited. The contract price may therefore have to include a great deal of overtime. This is expected to add about 5% to the cost.

Figure 15.1 shows how the estimate would be set out in an appropriate format, showing the changes in elemental rates. Figure 15.2 shows the same estimate using the usually preferred method of working out the elemental rate at the end. The estimated cost of £1,997,500 is slightly below the budget of £2,000,000 so is likely to be acceptable.

It is assumed that the question of professional fees is set out in a covering letter.

An example using BCIS data

The following example is worked out manually, although in practice the whole thing could be done on the computer using the BCIS *Online* service.

Example

A budget or early brief figure is required for a steel framed factory of 2,100 m^2 gross floor area located in Leicestershire. The tender date is expected to be in the fourth quarter of 1999.

	£
Mean cost/m^2 gross internal floor area of a steel framed factory adjusted to UK mean location at 1st quarter 1998 prices	341/m^2
Adjust for location Leicestershire × 0.93	317/m^2
Mean building price of factory in Leicestershire 2,100 m^2 at £317/m^2	£665,700
Allow for local pricing adjustment based on QS's knowledge of local market say × 0.95	£632,415
Allowance for inflation to fourth quarter 1999	
BCIS All-in Tender Price Index	
1st quarter 1998 = 141	
4th quarter 1999 = 156 (forecast)	
Adjustment $\dfrac{£632,415 \times 156}{141}$	£699,693
Approximate estimate at fourth quarter 1999	£700,000
Excludes external works, professional fees, and VAT	

Cost reductions

It may be that the estimated cost is higher than the client can afford, or more than the client wants to pay. It would be very dangerous to reduce the amount of the estimate for this reason, or on some vague grounds such as a general reduction in standard, although of course adjustments could be made for specific items such as (on the basis of the first example) cheaper lifts or a return to the cheaper floor finish.

However, if the total of the estimate is much too high it is best to face facts, and either reduce the size of the building or get the brief or the budget substantially modified. It is the easiest thing in the world to cut an estimate by 10% (or by any other percentage) under the influence of the client's pleadings and the designer's optimism.

PRELIMINARY ESTIMATE OF COST

PROJECT: Office Block, Midtown
DATE OF ESTIMATE: 12.02.99
ASSUMED DATE OF TENDER: Sept 1999
GROSS INTERNAL FLOOR AREA: 2,400 m^2

	£/m^2
BASIS OF ESTIMATE:	
Office Block, outer London (Sept 1998)	
1235 m^2	680.00
ADJUSTMENTS	
1. Market conditions	
BCIS Index. Add 10%	+ 68.00
	748.00
2. Size, number of storeys, etc.	
6 storeys in lieu of 4 – see notes	+ 6.00
	754.00
3. Specification level	
See notes. Wood block flooring £8	
Portland stone £35.75	+ 43.75
	797.75
4. Inclusions and exclusions	
See notes. Office partitions and decorations excluded	– 31.53
	766.22
5. Services	
See consultants and notes	+ 32.76
	798.98
6. Site and foundation conditions	
Piling (90 No 4 m long)	+ 16.87
	815.85
7. Other factors	
Tendering (see notes) add 2%	+ 16.32
	832.17
Professional fees	excluded
	832.17

TOTAL ESTIMATED COST: 2400 m^2 at £832.17/m^2 say £1,997,500

INCLUDED (and as above):

EXCLUDED (and as above):
Floor, ceiling and wall finishes in shops
All furnishings, fittings, blinds
Electric light fittings (except in public spaces)
VAT
Professional fees

Fig. 15.1 Preliminary estimate of cost (based on changes in elemental rates).

PRELIMINARY ESTIMATE OF COST

PROJECT: Office Block, Midtown
DATE OF ESTIMATE: 12.02.99
ASSUMED DATE OF TENDER: Sept 1999
GROSS INTERNAL FLOOR AREA: 2,400 m^2
BASIS OF ESTIMATE: Office Block, outer London (Sept 1998) £680.00/m^2

		£
NEW ESTIMATE: 2,400 m^2 at £680		1,632,000
ADJUSTMENTS		
1. Market conditions		
BCIS Index. Add 10%	+	163,200
		1,795,200
2. Size, number of storeys, etc.		
6 storeys in lieu of 4 – see notes (2,400 m^2 at £6)	+	14,400
		1,809,600
3. Specification level		
See notes. Wood block flooring £19,200		
Portland stone £85,800	+	105,000
		1,914,600
4. Inclusions and exclusions		
See notes. Office partitions and decorations excluded	−	75,700
		1,838,900
5. Services		
See consultants and notes 2,400 m^2 × 32.76/m^2	+	78,624
		1,917,524
6. Site and foundation conditions		
Piling (90 No 4 m long)	+	40,500
		1,958,024
7. Other factors		
Tendering (see notes) add 2%	+	39,160
		1,997,184
Professional fees		excluded
		1,997,184

TOTAL ESTIMATED COST: say £1,997,500 = £832.29/m^2

INCLUDED (and as above):

EXCLUDED (and as above):
Floor, ceiling and wall finishes in shops
All furnishings, fittings, blinds
Electric light fittings (except in public spaces)
VAT
Professional fees

Fig. 15.2 Preliminary estimate of cost (based on lump sum adjustments).

It must be emphasised that an overall 10% cut in costs means much more than a 10% cut in standards as much of the structural work, for example, cannot be reduced in cost. It must also be remembered, as stated earlier, that the estimate will carry the cost planner's signature, not those of the people who want to modify it. It can be assumed therefore that if the cost planner amends the estimate from its original figure there must be some tangible reason or reasons. These reasons need to be spelt out if a modified figure is presented.

Finally, mention must be made of the threat 'the job won't go ahead on this figure, do you want it to go ahead or not?'. This may be the response of a profit-oriented client, or an architect who does not want to lose a commission. Alternatively, it may be the attitude of a public body which knows that a major project will never get the go-ahead if its true cost is known beforehand, but will be very difficult to stop at a later stage when the enormity of its cost becomes apparent. Nobody ever seriously believed, for example, that the Sydney Opera House would only cost the figure of one or two million dollars which was first given. And closer to home, the British Library would have been unlikely to have received the go-ahead if an honest estimate had been given!

We are now well into the realms of professional ethics, about which people must make up their own minds in the light of the case before them, but whatever the ethical position the responsibility of the cost planner for whatever figure goes forward must be remembered.

To repeat what has been said twice already, the other parties will adopt a very low profile if things go wrong!

Data sources

The example of the two office blocks assumed that there was data available on a comparable scheme. In the later stages of estimating, when a cost plan is prepared on an elemental basis, an exact similarity of use between data source and proposed project is not vitally important – costs/m^2 of external walls, windows, doors, floors etc., need not vary much between a library and an office block. The costs/m^2 of floor area of the buildings themselves, however, will vary considerably because of the differing proportion of these items in them.

If there are no truly comparable schemes in the office it may be possible to get information from a professional colleague, or from the BCIS, or as a last resort to use published information.

When working from one's own information it is best to use one building as an example because the variables in design and tendering are known, but it is often helpful to supplement this with published information for similar buildings. As many examples as possible should be obtained to try and average out all the variable factors. Published information is always useful for checking and comparing an estimate prepared from one's own sources.

Mode of working

It may be objected that the above methods of working are far too crude. For instance, instead of guessing a percentage for an accelerated programme for the office block it would surely be better to compare the consequences of a normal programme and an accelerated one by preparing network diagrams and costing out the two sets of resources.

Well, of course it would, except that no detailed design exists for the building at this stage, and although it might be possible to study the effects in detail for a known building of fairly similar characteristics the answer would not be sufficiently valid to justify the amount of work involved (and in any case there was a 'market forces' component to the figure based on restricted tendering which such an exercise would not disclose). Also the workload to achieve this would be quite heavy; even if computer simulations are used they require a fairly considerable data input.

To repeat what was said at the beginning, what is being looked for is simply a 'ball-park' figure for a hypothetical development – ('How many noughts?') – and often there may not be enough firm information even to carry out the very modest range of adjustments set out above.

There will be plenty of opportunity to use more sophisticated techniques as soon as there are some tangible proposals to examine, and this process begins with the preparation of tentative sketch designs.

Summing-up

This is the stage where the big cost decisions are made, and design work should not start until the brief itself is right. There is little point in spending large sums of money meticulously – and successfully – cost planning and controlling a design which is not in fact a very good answer to the client's needs.

The preliminary estimates will almost certainly be based on the cost per square metre of floor area, preferably using a computer spreadsheet to jog the memory and make sure that nothing has been accidentally left out. It is usual practice to exclude VAT from the figures. The reports to the client must make clear what is included and what is excluded. Cost planners must be prepared to back their judgement if their estimates come out to a higher figure than the client is expecting.

Chapter 16
Cost Planning at Scheme Development Stage

Elemental estimates

As soon as the first sketch drawings are available an outline elemental estimate should be prepared in order to see whether the architect's initial solution is possible within the cost limits which have been set. This will be particularly important if the shape or design is rather unusual, or looks inherently expensive.

Like the preliminary estimate this is sometimes called a 'preliminary cost plan', but this is still misleading, since it is really only an approximate estimate, and it is probably the elemental format which leads to this terminology. The real cost plan will be prepared when this elemental estimate has been finalised, and it will be in a fresh format.

There are some advantages in using a 'short list' of consolidated elements for preparing the preliminary plan, as the only measurable items will be the main floor and wall areas and there is unlikely to be a specification as yet. For the purpose of this example a short list is assumed, and four main areas need to be measured, as follows:

- ground floor area;
- total floor area;
- area of external walls, including windows and doors;
- area of internal walls, including windows and doors.

Let us look at an example, based on the architect's first sketch design for the six-storey office block developed from the brief and estimate which were dealt with in Chapter 15. This is shown in Fig. 16.1.

Example

Gross internal floor area

Ground floor area	$30.00 \times 7.50 = 225.00$		
	$22.50 \times 7.50 = \underline{168.75}$		
	$\underline{393.75} \times 6$ floors	2,362.50	
Tank room	8.00×6.00	$\underline{48.00}$	
		$\underline{2,410.50}$ m^2	

213

External floor area

Main walls	$4 \times 30.00 \times 27.00$	3,240
Tank room	28.00×2.50	70
		3,310 m^2
Internal wall area, say		1,200 m^2

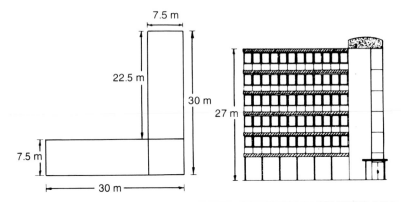

Fig. 16.1   Architect's first sketch design for provincial office block.

Already we can see that the ratio of external wall area to total internal floor area looks rather high (3310:2410 = 1.37 to 1), and we may feel that this is going to prove an expensive building. However, the elemental estimate will confirm or reject this theory better than any amount of conjecture. The best rates for pricing would be those obtained from an analysis of the outer London block, since we based the first estimate on the same project.

We must remember to add the 10% price increase and the 2% tendering differential which were included in the approximate estimate in Chapter 15, so as to keep the two estimates on the same basis, unless circumstances have changed in the interim. In which case, of course, the total of the estimate will have to be revised.

Again, a standard form should be used. The form shown in Fig. 16.2 is an example of the sort of thing required. Some of its features would vary according to the preferences of the individual office – for instance some people might prefer to show the preliminaries as a separate item, or to make a separate element of the frame. On the other hand, it would be possible to combine the finishes with the relevant structural items. However, the selected arrangement should be standard within the office so that the recorded information can be kept in a suitable manner.

The overall rates for elemental estimates can be obtained when doing the ordinary analysis and filed separately, and being large omnibus items there is perhaps less risk than usual in poaching prices from more than one job. With the rise of integrated computer systems this information can be stored and manipulated with far greater ease than was previously the case.

While the original practitioners of elemental cost planning believed in showing the elemental costs for each element (as in Fig. 16.2), it is more usual today to omit them, and just show the total costs for each.

Let us return to the particular example. The establishment of rates for the

```
┌──────────────────────────────────────────────────────────────────────────┐
│                        ELEMENTAL ESTIMATE No 1                             │
│                                                                            │
│  PROJECT: Office Block, Midtown                                            │
│  DATE OF ESTIMATE: 03.04.99                                                │
│  ASSUMED DATE OF TENDER: Sept 1999        Total cost (£)    Cost (£/m²)     │
│                                                             of floor area  │
│                                                             (2,411 m²)      │
│                                                                            │
│  GRD FLR AREA      (A)     394 m²                                          │
│  TOTAL FLR AREA    (B)     2,411 m²                                        │
│  EXTL WALL AREA    (C)     3,310 m²                                        │
│  INTL WALL AREA    (D)     1,200 m²                                        │
│                                                                            │
│  Work below lowest floor                                                   │
│    finish          (A)         394 × 321.00        126,474      52.45      │
│  Upper floors inc frame                                                    │
│    & stairs        (B–A)     2,017 × 123.50        249,100     103.32      │
│  Roof inc frame    (A)         394 × 164.50         64,813      26.88      │
│  External walls    (C)       3,310 × 212.00        701,720     291.04      │
│  Internal walls    (D)       1,200 ×  50.00         60,000      24.89      │
│  Floor finish      (B)       2,411 ×  44.00        106,084      44.00      │
│  Ceiling finish    (B)       2,411 ×  27.50         66,303      27.50      │
│  Wall finish       (C+2D)    5,710 ×  13.00         74,230      30.79      │
│  Decoration        (B)       2,411 ×   4.50         10,850       4.50      │
│  Fittings                                            2,750       1.14      │
│  Services                                          476,225     197.52      │
│  Drainage                                           19,000       7.88      │
│  Site works                                         27,500      11.41      │
│  Contingencies                                      30,000      12.44      │
│                                                  ─────────    ───────      │
│                                                  2,015,049     835.76      │
│  Professional fees                                     —          —        │
│                                                  ─────────    ───────      │
│                                                  2,015,049     835.76      │
│                                                                            │
│  (inclusions and exclusions as approximate estimate unless otherwise       │
│  stated)                                                                   │
│  VAT not included                                                          │
└──────────────────────────────────────────────────────────────────────────┘
```

Fig. 16.2 Preliminary cost plan.

various elements will have been complicated in this instance by the different proportion of unfinished shops compared to the outer London job.

A typical elemental rate calculation

The rate for floor finishes, for example, will have been obtained as follows:

Project used for comparison (outer London)

	£
Elemental rate for floor finishes on outer London project	27.90/m²
Add 10% increased costs and 2% tendering	3.35
	£31.25/m²

Calculation of floor finish elemental rate for areas excluding 'shell and core' shops.

		£
1,235 m² at £31.25 m²		38,594
Shops 25% = 309 m² at say £10.00		3,090
Balance 926 m²		£35,504

which divided by 926 gives £38.34/m² of gross internal floor area.

New provincial block 2,410 m²	£
16.67% shops = 402 m² at £10.00	4,020
83.33% remainder = 2,008 m² at £38.34	76,987
Total 2,410 m²	£81,007

which divided by 2,410 gives £33.61/m² of gross internal floor area.

	£33.61/m²
Add say 5/6 of £12.00 for more expensive floor finish in offices (as approximate estimate)	10.00
	£43.61/m²
Say	£44.00/m²

   This complicated calculation would have been much easier if the unit costs for the floor finishes had been available for the outer London job, and it would not have been necessary to use the risky proportion method.

   Alternative calculation using unit costs
   Unit rates taken from detailed analysis of outer London project, plus 10% inflation allowance and 2% tendering allowance.
   Quantities measured from architect's drawing of provincial job.

			£
Office areas	1688 m² at 34.25	=	57,814
Shops	338 m² at 8.95	=	3,025
Tank room	48 m² at 8.05	=	386
Entrance, staircase, toilets, etc.	337 m² at 45.20	=	15,232
			76,457
Add 10% increased costs and 2% tendering			9,175
			85,632
Add extra cost of better floor finish in offices	1688 m² at 12.00/m² =		20,256
	Total cost		£105,888
Elemental cost (divide by 2,411)			£43.92/m²
Say			£44.00/m²

Examination of alternatives

From an examination of the elemental estimate shown in Fig. 16.2 we can see that the scheme at £2,015,049 is likely to prove more expensive than the amount of the first approximate estimate (£1,997,500). Much of this increase is due to the slightly increased size of the building as designed (2,410 m² instead of 2,400 m²), but the cost/m² of the floor area is also increased from £832.17 to £835.76, and the estimated cost is now slightly over, instead of under, the budget of £2,000,000. While the increase is not very large (it could be got rid of by a small reduction in the level of specification), it does make us wonder whether this is the most economical solution to the design problem. At this point the figures in the right hand column (cost per square metre of floor area) come in very useful in enabling us to make comparisons with other contracts.

It certainly seems as though the external walls are costing a lot, which confirms our first thoughts on looking at the plan, but we now have figures to prove the matter. Although it would be practicable to build this scheme within the budget, it would be worth the architect's while to see whether an alternative shape would be possible.

The rather unsatisfactory appearance of the first elemental estimate should lead to the consideration of alternative schemes. It may not always be possible to improve wall/floor ratios very much, due to site restrictions or other unalterable factors. But let us suppose that in the case of the provincial office block the architect has been able to produce an acceptable alternative design (Fig. 16.3):

$$\text{Mean length of building} = \frac{(33\,\text{m} + 29\,\text{m})}{2} + \frac{(14\,\text{m} + 10\,\text{m})}{2} = 43\,\text{m}$$

$$\text{Ground floor area} = 43\,\text{m} \times 9\,\text{m} = 387\,\text{m}^2$$

$$\begin{aligned} \text{Total floor area } 6 \times 387\,\text{m}^2 &= 2{,}322\,\text{m}^2 \\ \text{plus tank room} \quad &\underline{\quad 48 \quad} \\ &2{,}370\,\text{m}^2 \end{aligned}$$

$$\text{Length of external walls} = 23.4\,\text{m} + 13\,\text{m} + 9\,\text{m} + 19.6\,\text{m} + 29\,\text{m} + 9\,\text{m} = 103\,\text{m}$$
$$\text{Area of external walls} \quad = 103\,\text{m} \times 27\,\text{m} = \quad 2{,}781\,\text{m}^2$$
$$\text{Add tank room walls as before} \quad \underline{\quad 48 \quad}$$
$$2{,}851\,\text{m}^2$$

Internal wall area say 1,200 m² as before.

The approximate cost for this scheme is shown in elemental estimate No 2 (Fig. 16.4). This is a much more hopeful design with an estimated cost of £1,918,550, slightly below the approximate estimate and the budget. The allowance for site works has been increased, as additional treatment to the front areas would be required owing to the front elevation being set back on the splay.

It should be realised that outline elemental estimates will not be completely accurate, although the comparisons obtained from them will be valid enough. Each of these breakdowns would only take a few hours of a senior assistant's time, and each hour spent at this stage is worth days spent on cost checks of minor elements later on in a frantic attempt to keep within budget. This is where the process of cost planning really pays for itself.

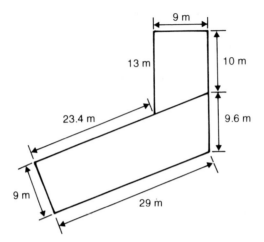

Fig. 16.3   Revised sketch plan for provincial office block.

Need for care

This is the place for a word of warning. Because we are dealing in such large figures at elemental estimate stage it is vital that all arithmetic, and if possible all measuring, should be checked. A silly mistake in calculating the principal areas would have serious consequences later on, and just because they seem (and are) simple, the elemental estimates must not be taken lightly. We should remember that 'everyone makes mistakes', particularly people working under pressure – checking will ensure that there are none. The same remarks of course apply to the preliminary estimate, which we dealt with in the previous chapter.

It is also important to remember that there are many other considerations in designing a building besides cost, and that the cost planner must not be tempted to try and usurp the architect's function when the shape and type of building is under consideration. The cost planner must be prepared to give the architect all possible assistance at this time, but should avoid making detailed suggestions about matters of planning and architecture which are entirely the architect's responsibility.

The cost plan

When the sketch drawings have been finalised and the budget (modified if necessary) has been accepted by the client it is now time to prepare the cost plan itself.

Although the establishment of the brief and the investigation of a satisfactory solution are being dealt with as two separate and consecutive functions, there may be a certain amount of iteration. Design investigation may suggest modifications to the brief, which in turn will need to be investigated. This is all to the good, and will probably result in improved performance. Even if it involves,

ELEMENTAL ESTIMATE No 2

PROJECT: Office Block, Midtown
DATE OF ESTIMATE: 10.04.99
ASSUMED DATE OF TENDER: Sept 1999

			Total cost (£)	Cost (£/m^2) of floor area (2,370 m^2)
GRD FLR AREA	(A)	387 m^2		
TOTAL FLR AREA	(B)	2,370 m^2		
EXTL WALL AREA	(C)	2,851 m^2		
INTL WALL AREA	(D)	1,200 m^2		
Work below lowest floor finish	(A)	387 × 321.00	124,227	52.42
Upper floors inc frame & stairs	(B–A)	1,983 × 123.50	244,901	103.33
Roof inc frame	(A)	387 × 164.50	63,662	26.86
External walls	(C)	2,851 × 212.00	604,412	255.02
Internal walls	(D)	1,200 × 50.00	60,000	25.31
Floor finish	(B)	2,370 × 44.00	104,280	44.00
Ceiling finish	(B)	2,370 × 27.50	65,175	27.50
Wall finish	(C + 2D)	5,251 × 13.00	68,263	28.80
Decoration	(B)	2,370 × 4.50	10,655	4.50
Fittings			2,750	1.16
Services			476,225	200.94
Drainage			19,000	8.02
Site works			45,000	18.99
Contingencies			30,000	12.66
			1,918,550	809.51
Professional fees			—	—
			1,918,550	809.51

(inclusions and exclusions as approximate estimate unless otherwise stated)
VAT not included

Fig. 16.4 Revised elemental estimate.

as it will, a good deal of abortive cost-planning work, the cost planner's work at this stage is relatively cheap compared to the potential benefits.

The cost control of design development, on the other hand, demands considerable resources, and should not be carried out until a satisfactory solution has been defined and agreed. As has already been pointed out, substantial iteration between the second stage (investigation of a solution) and the third stage (cost control of the development of design) of the cost planning process brings nothing but disadvantages.

The basic principle to be adopted is one of moving from the 'ball-park' estimating of outline proposals to the detailed costing of production drawings in a series of steps. The cost plan will probably be based upon the most recent elemental cost estimate for the project. It will, however, be developed with a full list of elements instead of the condensed list used so far, complete with an outline specification (agreed with the designer) for each of the elements.

This expensive document will form the basis for the system of design cost control. It should not be produced prior to approval of sketch design, since it might have to be done again a second (or third) time if the design changes. In any case the designer would find it difficult to make decisions on detailed specification matters while the design itself is still undecided. But it should not be left until the design is further developed. That may make it easier to do, but it will then become a record of what has been decided rather than a plan for controlling design – just a rather fancy cost estimate in fact.

The cost plan may take many forms, but all the different systems have much common ground.

Specification information in the cost plan

Opinions differ on how detailed the specification information should be when the cost plan is prepared and this is a suitable place to discuss the issues involved.

The first approach involves giving a fully detailed specification in the cost plan:

- The cost planner/quantity surveyor will try and get as much detailed information as possible from the architect before preparing the cost plan and will fill in any blanks with typical specifications within the required cost range.
- When the cost plan is complete the architect will be supplied with a detailed list showing the type of construction and finish on which the cost is based.
- The architect knows that although anything in the list may be altered at will, any such deviation may have an effect on the planned cost.
- The cost planner will feel that this specification list gives some protection in the possible event of an architect failing to design within the agreed limits, since these limits are written into the cost plan and are not just assumed.
- However, this procedure is open to the objection that the cost planner appears to be taking the responsibility for design out of the architect's hands.

The alternative way of dealing with this problem is to leave the responsibility for design entirely where it belongs, with the architect:

- When the cost plan is prepared the architect's ideas on specification are incorporated into it, but if these are not forthcoming the cost planner does not provide them.
- Instead, a target cost for the element is devised, based on standards of cost performance achieved elsewhere, and it is left to the architect to design within that cost.
- If the element, when designed, costs more than the sum of money allocated to it then it must either be redesigned or the elemental cost increased at the expense of other elements.

That the theory behind this second approach is sound few will deny, but the first alternative is the one which was always preferred by most practising cost planners, and today is almost universally adopted.

Apart from the factor already mentioned (of protection against an uncooperative architect) it is said that most architects prefer the discipline of a specification to the provision of a cost figure for the element which is unrelated to any specification and may necessitate several attempts at design before it is achieved. The 'detailed specification' method also has the big advantage that if the architect works to the agreed specification then subsequent cost checks will be unnecessary.

Elemental cost studies

Cost research or cost studies may be required when the architect is considering design alternatives. It is at this stage that cost models such as those described earlier will be most effective. However, at the present time these studies would normally be done by means of approximate quantities, about which sufficient has already been said.

Cost studies, if done properly, are expensive in professional time, and therefore cannot usually be undertaken in respect of the majority of items in the project. They should therefore be reserved for comparisons of important components, or for projects which are themselves unusual and for which ordinary elemental cost data are of little use.

Large-scale cost studies are inappropriate where the component is not a significant part of the cost structure, or where the difference in cost between the alternatives is obviously going to have little effect on overall costs. Where there is little cost difference between one solution and another the decision should be taken on grounds other than cost and not on the possibility of a marginal saving.

If the contract is to be awarded in the traditional manner with competitive tendering on BQs no quantity surveyor can possibly forecast which of two alternatives with a marginal cost difference would actually be priced the cheaper by the (unknown) successful contractor. In connection with this the cost planner must remember the natural tendency of contractors to play safe when pricing work involving unknown or experimental materials and techniques.

Another point to bear in mind in connection with cost studies is that all constructional systems have certain optimum conditions (of loading, span, etc.) in which they are at their most economical. There is usually a reasonable spread of conditions on either side of the optimum in which the system remains reasonably economical, but once the conditions get outside this range costs will start to rise steeply. Even though the particular system may still be perfectly feasible in these unsuitable conditions, it is likely that there will be a cheaper and better solution.

It is also worth noting that a high standard of fire resistance, thermal insulation or sound insulation normally costs money, and that it is usually wasteful to employ methods of construction or materials which offer these advantages if they are not required.

We will now look at some of the principal elements in a building, and the particular aspects of design costs which affect each.

Foundations

The type of foundations required will normally be influenced by the type of structure which they are to carry; obviously a structure which imposes heavy point loads cannot be carried on conventional strip foundations.

Piling is almost always a very expensive solution and is normally employed only where conventional foundations are impossible because of the depth at which a bearing formation exists, combined sometimes with the waterlogged nature of the ground. Different systems of piling have particular advantages and disadvantages, but the specialist piling firms (most of whom offer several systems) are usually more than ready to offer their services in connection with early cost investigations.

As previously noted the cost of the piling itself will usually be a complete extra, because the complex of pile caps and ground beams associated with this form of construction is often as expensive as conventional foundations.

Frame

Cost studies may be necessary to determine the type of frame to be used or indeed whether there should be a frame at all.

External walls of habitable buildings need certain qualities of weather resistance and heat insulation, and such a wall capable of bearing quite substantial loads costs no more than a non-loadbearing wall with similar weathering and insulation performance.

In conditions where normal external walling would be capable of carrying the weight of floors and roof the cost of a frame is a complete extra, and it follows that for small buildings of not more than three storeys a frame is likely to be an expensive solution, even though it may be desirable for other reasons. A frame for small buildings only becomes economically justified where it is necessary to have a very high proportion of the external walls glazed, or in 'big sheds' such as warehouses and the like where the walls do not need good insulation or aesthetic qualities and cheap non-structural materials such as profiled cladding can be used.

A frame, if required, is likely to be of steel or reinforced concrete, with light alloys and timber as recent and rather specialised competitors. As with other comparisons between materials, if either steel or reinforced concrete held all the cost advantages the other would long since have gone off the market.

For multi-storey construction there is a general tendency for reinforced concrete to be cheaper in normal conditions of loading, although the difference may often be small. The structural steelwork industry is continually improving its techniques of design and construction, particularly when the cost competition from reinforced concrete becomes severe.

Although steel is inherently cheaper than reinforced concrete for most normal applications, the saving can sometimes be wiped out by the measures necessary to give the structure an adequate fire resistance. This is almost certain to be the case if in-situ concrete cladding is required to beams and columns, so that the type and standard of fire resistance is therefore an important factor in any such cost comparisons.

Steel therefore gives its best comparative cost performance as a frame to single-storey buildings or in roof framings where fire resistance is not normally required by the building regulations and where the steel may be used without any protection other than paint. For such applications reinforced concrete is not usually competitive, except in some proprietary systems for constructing factories or sheds out of standard precast units.

Steel frames have advantages over in-situ frames in site programming, and some contractors (particularly small or medium sized firms) prefer working with a steel frame to erecting a reinforced concrete structure of their own, and this preference would be reflected both in tendering enthusiasm and in completion time for the project.

In designing cost-effective steel frames, attempting to save weight of steel can be counterproductive if the result is increased complexity – by using compound welded members, for instance, in lieu of simple sections. A steel fabrication firm's advice should be sought as to the point where savings are swallowed up by fabrication costs. Most constructional steel firms and specialist reinforced concrete contractors are very helpful in giving cost information for a new project.

If a consulting structural engineer is appointed for the project the cost planner would obviously work with the consultant in providing cost information. As alternative frame designs may affect other elements, the structural engineer may not be able to ascertain the full cost implications without assistance.

As well as a choice of materials, frame design will involve a choice of frame shapes and spacings. The most uneconomical solution will occur where heavily loaded beams have to span long distances, especially if they are also restricted in depth. The spacing of columns and beams will have an important effect on cost, particularly if the columns are expressed on the elevations and covered with expensive cladding.

If the frames are spaced too closely the savings in sections will not pay for the additional frames, since although frames spaced at 2.5 m centres will each be carrying two-thirds of the weight of frames spaced at 3.75 m centres, they will cost much more than two-thirds of the latter. This is because the choice of section for a beam is affected by its own weight and span as well as by the load it carries, and the span of the beams will not have been changed (see Fig. 16.5).

Since the most economical spacing will depend upon the span and upon the loading it is necessary to consider each case individually. For normal floor loadings, however, it is unlikely that spacing as close as 2.5 m will give the most economical solution, while spacings much in excess of 5 m will begin to produce additional costs on the floors and roof owing to their increased span, even though the frame design itself may still be economical.

Staircases

Since the structure of a concrete staircase represents less than a third of its total cost (the remainder being caused by finishes and by balustrades) and since the

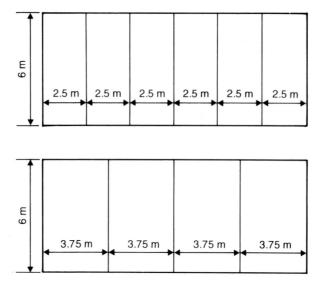

Fig. 16.5 Alternative grid layouts.

whole elemental cost is only a minor one it is doubtful whether the staircases of a building are the most fruitful field for cost studies.

However, as well as the staircase and its finishings, balustrades, etc., the true total cost of a staircase would include:

- the surrounding walls of the staircase together with foundations,
- windows,
- wall finishes,
- doors,
- roof, etc.,

and would be quite considerable. The most rewarding approach is therefore likely to be a reduction in number of staircases rather than in the details of their design.

Upper floors

In contrast to staircases, upper floors are an important element and require comparatively little work in the evaluation of alternative designs, which may therefore be well worth doing. As with some other elements, it is difficult to arrive at true cost comparisons without considering the 'frame' element as well.

Roofs

The architect may require cost studies of alternative roofs to which most of the preceding remarks apply. In addition, the comparative costs of flat roofs and pitched roofs may be involved. These will depend more upon the level of the two specifications than upon the basic difference in roof type. However, for medium to large spans a satisfactory pitched roof is likely to be rather cheaper than a flat roof of comparable quality, partly because of the simplicity of spanning large areas with roof trusses rather than deep beams. It is also often possible to use the resulting roof space as part of the accommodation area.

Pitched roofs also tend to be more durable than flat roofs, which are sensitive to poor workmanship and poor design detailing unless very expensively built. Bad maintenance experiences with cheap flat roofs has led to the avoidance of these wherever possible.

Rooflights

Rooflighting is not normally the most economical method of lighting rooms where windows are a possible alternative, and individual domelights or small lantern lights are probably the least satisfactory from a cost point of view. However, there are many reasons why the architect may choose this means of lighting, for instance openable lights for natural ventilation or trickle vents, and the element is not important enough for this to affect the total building cost very considerably.

On some current buildings steeply pitched roofs effectively form the walls of upper storeys, and these require the same level of fenestration as a wall. This is one of the areas where the theory of elemental cost planning breaks down, and it will almost certainly be necessary to consider the walls and the roof together for cost planning purposes.

External walls

This may be the most important structural element, particularly in a multi-storey building, and being one in which a tremendous range of constructional methods and finishes may be used is a suitable element for cost studies.

In the type of building where the external walls comprise a series of repeating panels it will be sufficient to study a single panel in detail. It will often be difficult to consider this element in isolation from windows, internal finishes and frame.

Note that ordinary brick or concrete block walls finished with facing bricks externally are likely to be far cheaper than any other construction of comparable performance and durability. Therefore a considerable area of plain walling on a building may enable smaller areas of luxury construction or finish to be used while still keeping the overall elemental cost quite reasonable.

Internal walls and partitions

The comparative costs of traditional partitioning methods are well enough known for cost studies to be unnecessary, but if it is proposed to use a special type of partition on a large scale then a cost investigation would be worthwhile. Comparative costs of small areas of special partition or glazed screens, on the other hand, would not have much effect on the total building cost and would not usually be worth investigating.

If the BCIS or similar list of elements is being used it is important to remember that the cost of finishings to traditional partitions is included in other elements, but the complete cost of self-finished proprietary partitioning is included in the 'partitions' element, so that in this instance yet again the elemental costs will not be strictly comparable.

Windows

These are often a major element of modern buildings and have a very large cost range.

Cost differences will normally be due to performance requirements rather than the actual materials used in manufacture; standard metal, UPVC or timber windows for housing schemes are competitive in cost as are high-class metal or hardwood windows for prestige buildings, but the latter category may be four or five times dearer than the former.

Apart from the actual material and section of framing the cost of windows is substantially affected by performance, particularly as regards double glazing and weather stripping, but there are also factors affecting individual windows or groups of windows rather than the fenestration of the building as a whole:

- *Size of window.* Small windows tend to be high in cost per square metre, because of both the greater intricacy of the window itself and the cost of forming the opening in the wall; these two costs vary according to perimeter rather than area.

- *Size of panes.* Very small panes increase the cost per square metre, as do panes which are so large that it is necessary to glaze them with stout float glass instead of sheet glass. It is unlikely that any saving on window frame or glazing bars would counterbalance the cost of plate glass unless an extremely expensive type of window was being used.

- *Opening lights.* This is probably one of the most important factors in window cost, particularly where a high standard of weather resistance is required. The heavy comparative cost of opening portions of windows (as against areas of fixed glazing) makes it very difficult to compare window costs on an overall square metre basis unless the pattern of opening and fixed areas is very similar. Unfortunately it is not even possible to compare the costs of opening portions on a square metre basis as so much of the cost (hinges or pivots, fasteners, framing to angles) varies according to the number of sashes rather than their size.

- *Decoration and glazing.* Some types of window come to the site ready glazed and self-finished, others require to be glazed on site and painted.

- *Special types of glass.* For example, solar reflective glass or the laminated glass required by Building Regulations in some windows or screens.

Because of the above factors it is often necessary for any cost studies of windows to require individual consideration of all the window types in the building, rather than being confined to a typical window and the results being applied to the remainder on a square metre basis.

Doors

Except on buildings which are divided into a large number of very small rooms, such as hostels or flats, or where non-standard sizes or fancy ironmongery are used, the doors are not usually a significant element. Because of this, because they are one of the components most subject to heavy mechanical wear, and because they are also one of the most noticeable features of the building it is probably a mistake to be excessively concerned with cost when designing this element.

Floor, wall and ceiling finishes and decorations

Cost studies of these elements are likely to be comparatively simple, and where large areas are involved are certainly worth doing.

Engineering services

On most buildings this is one of the most important elements. It is essential that cost studies be carried out, and these will be fairly meaningless unless 'whole life' investigations into running costs, performance and updating/replacement are taken into account.

Most cost planners are unlikely to have access to the data on which such studies should be based and will need to work closely with specialist engineers, preferably with those who are going to be responsible for the design.

Joinery fittings

Of all minor building components single joinery fittings involve the most work in arriving at a cost, and so cost studies of such fittings should not usually be attempted. Where a fitting has been designed for repetition (such as a bedroom fitting for a large hostel or a typical bench to be used as a basis for furnishing a set of laboratories) the position is of course different.

Cost studies generally

We have seen many ways in which cost studies can contribute to the design of a building. However, it is important to realise that it would not be economically possible to employ more than a few of them on any particular building, as a complete cost study of every element, major and minor, would not produce savings commensurate with the amount of time expended. The only occasion where such a course might be practicable is where a large number of similar buildings are to be erected, as for instance a standard house design for a large housing authority.

Preparation of the cost plan

We now come back to considering the preparation of the cost plan itself. This may take many forms, but all the different systems have much common ground.

The plan will almost certainly be prepared by elements and the cost will be expressed per square metre of floor area even if it is calculated in a different manner, in order to be able to make comparisons between other schemes of different size on a common basis. If there is a cost target it will by now have been set up with some degree of finality and the architect will want to go ahead with the working drawings.

Since cost planning involves cost control the architect must design in accordance with the cost plan, which must therefore be available before the working drawings have proceeded very far. What do we need in order to prepare such a plan?

- A drawing and a standard of specification to which the cost plan can be related.
- A cost analysis of a comparable project.

It is not necessary that the 'comparable project' should be for a similar use as long as there is some reasonable compatibility as buildings. For instance, a police station and a health clinic are both public buildings with fairly similar storey heights and are divided into both small and large rooms. The elemental analysis of a police station could therefore be used to prepare the cost plan for a health clinic if nothing better was available, whereas between say a church and a block of multi-storey flats there is no resemblance whatever.

The cost plan must be prepared to a standard format, for reasons that have already been emphasised in connection with analyses and preliminary estimates. The form should show:

- the chosen list of elements;
- preliminaries and insurances (if separate);
- contingency sum;
- professional fees (if to be included);
- the cost index value which is being used.

It is common practice in some offices to allow a 'design margin' of between 1% and 5% as a design contingency in addition to the contract contingency, and this margin may come in useful during the cost check when it can be absorbed into any element that needs it.

Although specialist computer software is available for cost planning work, many cost planners prefer to use an ordinary spreadsheet package adapted to the practices of their own particular office. Whichever alternative is used, the software should guide the user through the process of preparing the cost plan, making sure that all necessary adjustments are made and that nothing is forgotten. Prior to the large-scale adoption of computers for the purpose much the same control was obtained by the use of standard manual forms and procedures. The first page of a typical manual form is shown in Fig. 16.6 and the accompanying specification in Fig. 16.7.

COST PLAN

PROJECT: Office Block, Midtown
DATE OF COST PLAN: 24.04.99
ASSUMED DATE OF TENDER: Sept 1999
TOTAL INTERNAL FLOOR AREA: 2,390 m²

Note: **This cost plan is based upon the attached outline specification, and both documents should be read together**

	Unit quantity	Unit cost (£)	£	Total cost (£)	Elemental cost/m² (£)
1. WORK BELOW LOWEST FLOOR FINISH					
Grd Flr Area	390 m²	321.00		125,190	52.38
2. STRUCTURAL FRAME					
	2,390 m²			125,600	52.55
3. UPPER FLOORS					
225 mm Hollow pot	386 m²	60.00	23,160		
150 mm in-situ RC	1,585 m²	41.00	64,985		
			88,145	88,145	36.88
4. STAIRCASES					
RC Staircases					
1 No 25 m rise	25 m	1225.00	30,625		
1 No secondary					
21.5 m rise	21.5 m	900.00	19,350		
			49,975	49,975	20.91
			continued 1.	388,910	162.72

Fig. 16.6   Front sheet of typical cost plan.

Method of relating elemental costs in proposed project to analysed example

There are two ways of doing this, either by using unit costs of the elements or by proportion. In spite of what may be thought to the contrary, these are really two ways of doing much the same thing. In both methods the unit quantity of the

---

OUTLINE SPECIFICATION
in connection with cost plan
for
Office Block, Midtown

Date: 24.04.99

WORK BELOW LOWEST FLOOR FINISH

Foundations to walls and/or columns	In-situ concrete piles average 4 m long (approx 90 No), reinforced concrete pile caps and ground beams
Basements, walkway ducts, etc.	Semi-basement 1.5 m deep approx. 8 m × 4 m for boilers Also lift pits approx 0.75 m deep in waterproofed concrete
Rising walls	Clay common bricks in cement mortar built off ground beams (approx 0.75 m high generally to dpc) Reinforced concrete walls (0.3 m thick) around stair well and lift pits
Ground floor slab	Generally 150 mm slab reinforced with fabric on building paper and hardcore Note thickness of hardcore to make up levels (0.60 m average)
Dpcs and membranes	Lead-cored bituminous dpc in walls 3-coat Synthaprufe or similar on gf slab 2-coat asphalt tanking to semi-basement

STRUCTURAL FRAME

Generally	Reinforced concrete beams (at 3.50 m c/c) and columns, similar to outer London house

Fig. 16.7   Specification notes to accompany cost plan.

element (e.g. area of external walls) has to be measured from the sketch drawings of the proposed project.

Using the first, or 'unit cost' method:

- This quantity is priced out at the elemental unit rate per square metre for the same element obtained from the comparable analysis, where this is applicable, but if specifications are different then approximate quantities are more likely to be used.
- The approximate quantities will give the total cost of the element, and this total is then divided by the floor area of the proposed project to give the elemental price per square metre.

When the second, or 'proportion' method is used:

- The ratio of unit area to floor area in the proposed project is compared to the corresponding ratio in the analysed example.
- The elemental cost per square metre is then adjusted in proportion to the difference in the two ratios.
- With this method one is not concerned with the actual unit cost (either in total or per square metre) at all, although the total cost for the element will be required for comparison purposes at cost check stage.

What are the advantages and disadvantages of the two methods? The proportion method is the logical choice for use where the elemental cost per square metre is regarded as an index of performance rather than the result of a specification, and it tended to be preferred by central government departments in the early days of cost planning.

However, the unit cost method is almost exclusively used today. It is more in line with the traditional quantity surveying approach and most cost planners would feel less likely to make mistakes when using it, particularly if the workings become complicated. The unit cost method is more easily related to approximate quantities either at cost plan or cost check stage, and we must remember that at one or other of these stages fairly detailed quantities have got to be introduced.

In order to illustrate the systems in use, two examples are worked out below using both methods. It will be seen that either method will allow differences in quality as well as quantity to be adjusted.

Example 1

Total floor area 5,000 m^2
External wall area 4,000 m^2
Wall:floor ratio 0.80
Cost of external wall per m^2 of floor area (elemental cost) = £75.00
Cost of external wall per m^2 of wall (unit cost) = £93.75

Proposed project
Total floor area 4,800 m^2
External wall area 3,600 m^2

Elemental cost of external walls on proposed project

Unit cost method

3,600 m^2 of wall at £93.75 = £337,500

£337,500 ÷ 4,800 = £70,312

Total cost 4,800 × £70,312 = £337,500

Proportion method

Wall:floor ratio $= \dfrac{3,600}{4,800}$

$£75.00 \times \dfrac{0.75}{0.80} = £70,312$

Example 2

Total floor area 5,000 m²
Internal doors 35 no.
Cost of doors per m² of floor area (elemental cost) = £2.10
Cost of doors each (unit cost) £300

Proposed project
Total floor area 4,800 m²
Internal doors 50 no.
Cost of doors each £330 (better quality)

Elemental cost of internal doors on proposed project

Unit cost method

50 doors at £330 = £16,500

£16,500 ÷ 4,800 = £3.44

Total cost 4,800 m² × £3.44 = £16,500

Proportion method

$$£2.00 \times \overbrace{\frac{5,000}{35}}^{\substack{\text{Quantity}\\\text{adjustment}}} \times \frac{50}{4,800} \times \overbrace{\frac{33}{30}}^{\substack{\text{Quality}\\\text{adjustment}}} = £3.44$$

Sources of data

If the analysed example will not yield all the necessary information for preparing the cost plan this information must be obtained elsewhere.

One approach is to take a price from another analysis. This is potentially dangerous; we have seen how a priced BQ will probably contain compensating errors and that these will be incorporated in the analysis. However, if we take items from more than one analysis we may get cumulative instead of compensating errors.

It would be possible (indeed tempting) to take a number of elements from different analyses, all of which are priced too low, and this would make the total of the cost plan quite unrealistic. This cannot happen where a single analysis is used throughout.

If prices have to be obtained from other analyses it is wise to collect as many examples of the same item as possible so that any errors are likely to cancel out in the average so obtained.

The alternative is to use approximate quantities. These are likely to be very approximate indeed (remember that we have no working drawings) and herein lies the danger; it is much easier to leave things out than to put extraneous items in. Therefore when using approximate quantities the cost planner must be very careful to include all the items which are included in the description of the element (e.g. skirtings and their screeds with 'floor finishes' or rainwater pipes with 'roofs').

It is also important to include the percentages for preliminaries, insurances or contingencies if these are included in the elemental rates.

Needless to say, the quantities must be priced at the same level of market rates as the cost plan itself, although the rates will be higher than ordinary BQ prices because of the need to include the cost of secondary work which would be measured separately in a full BQ.

The approximate quantities themselves will be similar to those normally employed by quantity surveyors for estimating purposes except that the elements must be kept separate. So, for example, the plastering on the underside of a concrete roof slab could not be included in the overall rate for the roof but would be worked out separately as part of 'ceiling finishes'.

In some offices it is the custom to keep a complete schedule of prices for use when pricing approximate quantities. This certainly gives a consistent level of pricing, but it is doubtful whether any but the largest organisations would find the preparation and maintenance of such a list to be an economic proposition.

Examples of use of information from analysed building

Analysis
Total floor area 2,000 m^2
Element: internal partitions
Elemental cost: £27.00 m^2

	£
1,400 m^2 of 75 mm breeze partition at £13.00	18,200
900 m^2 of 225 mm and 300 mm brick walls at £33.00	29,700
120 m^2 of glazed partitions at £50.80	6,096
2,420	53,996

Average unit cost = £53,996 ÷ 2,420 = £22.32 m^2

Proposed project
Total floor area 3,100 m^2
Total area of partitions (measured roughly from drawings) 3,500 m^2

If we do not know the type of partition required, or if the proportions of various types are likely to be similar to those of the analysed example, we can use the average unit rate or work by the proportion method:

3,500 m^2 at £22.32 = £78,120 total cost divided by 3,100 =
£25.20/m^2 elemental cost

But if even at this stage we can see that the proportions of the various sorts of partition are quite different to the analysed example we must use the individual unit cost as shown below.

		£
20% 75 mm breeze partition = 700 m^2 at £13.00		9,100
10% 225 mm brick wall = 350 m^2 at £33.00		11,500
60% glazed partition (cheaper type) = 2,100 m^2 at £45.00		94,500
10% timber stud partition (no rate for this: price built up by approximate quantities) = 350 m^2 at £23.00		8,050
		123,200

£123,200 ÷ 3,100 m^2 = £39.74 elemental cost

When severely altering the proportion of items like this it is necessary to watch out for any freak rates. For instance, it is possible that the rates for glazed partitions in the analysed example may have been much too low, and being a very small part of the whole job this will not have had much effect on the total price. On the proposed project, however, the proportion of glazed partition has been increased from 10% to 60% and such an error would be much more serious. Again this emphasises the point which has been made before, that the smaller items in the analysis may well have the least reliable prices.

Elemental costs

By whatever means the total cost of the element has been obtained it should always be expressed per square metre of floor area, for the following two reasons:

- For comparison of reasonableness with other buildings. If the partitions (for instance) are costing far more per square metre of floor area than on other buildings of the same type it is worthwhile investigating the reasons. These may be perfectly sound; the cost may include built-in cupboards, there may be a greater degree of insulation required between rooms, but it is also possible that there is inefficient planning of the accommodation and this would also be reflected in other elements.
- The elemental price per square metre enables the importance of any extravagance or economy to be judged in a way that is not possible when considering unit prices only.

Comparison of elemental costs

Let us consider the unit costs for three different specifications of internal doors:

- Hardboard faced skeleton framed flush doors hung on steel butts to softwood frames, no architraves. Unit cost £45.00 per m^2.
- Plywood faced semi-solid core flush doors hung on steel butts to softwood frames with softwood architraves. Unit cost £82.50 per m^2.
- Mahogany veneered plywood faced solid core flush doors hung on solid drawn brass butts to mahogany frames with mahogany architraves. Unit cost £175.00 per m^2.

From a comparison of the unit costs it would seem that the third specification is prohibitively expensive for normal work compared with the more usual second specification. Yet to adopt the better specification in a typical school of 3,000 m^2 with 70 internal doors would only increase the elemental cost per square metre by about £2.16/m^2, and to use the completely inadequate first specification would only save £0.88/m^2. Considering that the total cost per square metre of the school would be around £600, these increases and decreases are quite insignificant.

On the other hand, a marginal difference in some other elements, such as the particular choice of hardwood for wood block flooring, which may only alter the unit price by 10%, will have as great an effect upon the cost of the school per square metre as a 100% difference in door prices. It will soon become obvious to the cost planner which items need the most time spent on them at cost plan stage – in other words, which elements have the greatest cost significance.

It would be worthwhile to use approximate quantities for a major element such as external walling where a difference of one or two pounds in the unit cost per square metre would have an important effect on total cost, and to check the cost plan very closely when drawings subsequently become available. Such elements as internal doors or ironmongery on the other hand can be dealt with much more rapidly.

Presentation of the cost plan

The cost plan should be presented as neatly as possible; it will almost certainly be going to the client and should reflect well on the firm which has prepared it. To what extent expense should be incurred in order to give the plan an impressive appearance is a matter for individual preference, but if the document has been prepared by computer a high-quality (preferably colour) printer should be used for the purpose.

Another matter for personal decision is the extent to which the detailed build-up of the plan should be made available to the architect. Whether this should be done in all cases will depend upon the degree of mutual confidence which exists and also upon the available facilities for copying a large quantity of rough working.

With the cost plan should go the specification, if any, upon which it is based; this will necessarily be more in an outline form than the type of specification which will eventually be embodied in the BQ. It may either be included as part of the cost plan itself or be prepared as a separate document. A typical example was shown in Fig. 16.7. If a separate document is used it is important that there should be a space for each specific item so that it can be quickly seen whether everything has been dealt with or not. The specification should preferably be filled in by the cost planner in consultation with the architects, rather than being sent to the architects for them to fill in by themselves.

Summing-up

Early estimates of the cost of the scheme will probably be done using a short-list of some half-dozen principal elements only. But because such large figures are being dealt with it is even more vital than usual that all arithmetic, and if possible all measuring, should be checked.

The cost control of design development demands considerable resources and should not be carried out until a satisfactory solution has been defined and agreed. Substantial iteration between the second stage (investigation of a solution) and the third stage (cost control of the development of design) of the cost planning process brings nothing but disadvantages. The basic principle to be adopted is one of moving from the 'ball-park' estimating of outline proposals to the detailed costing of production drawings in a series of steps.

The cost plan will probably be based upon the most recent outline elemental cost estimate, but developed in as much detail as a full elemental cost analysis complete with outline specification (agreed with the designer) for each of the elements. It is generally preferred that the cost plan should incorporate the specification on which it is based. Although elemental estimates can be based on approximate quantities or proportion methods, the former is usually preferred today. Cost studies of individual elements are generally expensive, and should be limited to elements of major cost significance or to elements, such as floor finishes, where not much work is involved.

Stage 3
Controlling the Cost

Chapter 17
Building Resources and Costs

Now that the cost plan has been approved and the working drawings are in hand we need to consider how the building is actually going to be built.

The construction industry

Developers normally need to involve themselves with the building industry in order to get their building work undertaken. The building industry and the civil engineering industry together are often referred to as the construction industry, and many firms (certainly most of the large ones) operate in both sectors, even though their staff and organisations will probably be separate.

It is, however, very difficult to separate the two halves of the construction industry statistically, because of the overlap which occurs. For instance, the construction of the foundations of very large buildings could almost be classed as civil engineering (and you will find it so referenced in your library, if this operates the Dewey or UDC system), and there may be quite a lot of building work on some civil engineering projects.

A further sub-division of the building industry is into housing and other work, many main contractors having divisions which specialise in one or the other.

The building industry in the UK has changed very considerably over the last 30 years, in two ways. First, most firms of general building contractors now undertake only a small proportion of their turnover using their own directly employed operatives, the majority of their work being out-sourced to specialist or trade sub-contractors. The general contractor's role has changed from being primarily a provider of resources into being a provider of management and financial services. Very often the supervisory and administrative staff and a handful of labourers will constitute the entire site workforce of the general contractor.

In the second case, between the end of the Second World War and the late 1970s the building industry was geared largely to serving central and local government departments as its direct clients. Because of changes in the country's economic structure this is no longer the case, as is shown in Fig. 17.1. In much of the country, and particularly in the south-east, the industry has had to accustom itself to the different norms of private enterprise clients, more interested in results than in procedures.

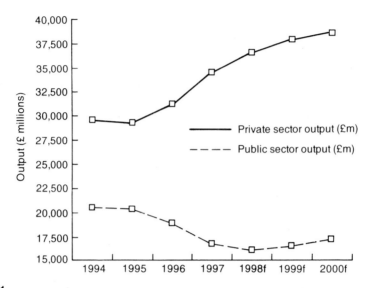

Fig. 17.1

Nature of the construction industry

The UK construction industry is very large, with a turnover in 1996 of around £55 billion according to the Stationery Office publication *General Economic Data 1998*. This represented 5.2% of the gross domestic product.

The industry consists mainly of small firms. Only 87 out of nearly 165,000 firms (0.05%) employ more than 600 people directly. However, these few large firms undertake 15% of all construction in Great Britain, although they usually employ smaller firms to do much of the actual work.

Because it operates with a relatively small investment in fixed assets in relation to its large turnover, the industry is very flexible and is able to accommodate itself to major changes in work load much more easily than industries which are plant-based. Its workforce can be reduced in hard times without the sort of political crisis caused by the closure of a car factory, and perhaps because of this it has often been used as an economic regulator by government. Similarly, it can expand its efforts rapidly in order to meet demand without requiring major investment – for instance, in the two years 1987–1988 the building industry was able to increase its output by one-fifth.

Problems of changes in demand

Nevertheless, violent changes in demand do harm the industry. Workers who are laid off in difficult times do not always return to the industry afterwards, and the problem is compounded by the need to train skilled workers. Firms say they cannot afford to do this when times are hard, and in boom times they are usually too busy to spare the time. It is also difficult to keep a balanced management team of the right size when demand fluctuates wildly, and as stated above management expertise is now probably the general contractor's main stock-in-trade.

A further problem concerns the supply of materials, because material and component manufacturers are plant-based and find it more difficult to respond rapidly to changes in demand than the construction industry itself. If a boom is regional in character this problem can be mitigated by importing materials and components from elsewhere in the UK or Europe, but for bulk materials in particular (where transportation forms a large part of their cost) this can often be an expensive alternative.

It is therefore very important for the cost planner to look at the general and local economic situations when forecasting costs. For example, a look at Fig. 14.1 on page 192 will show that for a few years after the start of the major slump in building in the early 1980s tender prices remained almost static, although official costs of labour and material inputs continued to show an increase. In the $4\frac{1}{2}$ years between the third quarter of 1980 and the first quarter of 1985 tender prices increased by less than 1%, but the official index of building costs rose by no less than 38% in the corresponding period! The same thing happened even more violently between 1990 and 1996.

This imbalance between tender prices and building costs was partly due, of course, to tenderers' willingness to take a cut in profits during difficult times. But it was mainly due to the fact that real costs did not increase, whatever the official figures might have said.

Theoretical inflationary increases were totally counterbalanced because:

- Labour could be easily obtained and no longer had to be bribed to work for one employer rather than another.
- With workers fearing the sack if they did not perform, productivity tended to improve.
- Under the stress of competition materials arrived when they were supposed to, and merchants started to offer discounts instead of requiring extra payments for prompt delivery.
- For the above reasons management of projects and the meeting of deadlines became much easier, and it was no longer necessary to make a substantial allowance in tenders for risk.

Although the above situation occurred on a national scale it is important to look out for similar supply-and-demand imbalances occurring regionally or even locally. Much of the building industry consists of smaller firms operating within a limited radius, and labour is only mobile to a limited degree.

Costs and prices

There are two ways in which building costs can be estimated or analysed, based on:

- the prices charged for the finished building or parts thereof or
- the cost of the resources required to create them.

Most of the cost planning and cost control procedures traditionally used by quantity surveyors on behalf of the client or design team have depended upon 'prices' for finished work-in-place, obtained from BQs or elsewhere, because this is what the client will actually have to pay. However, these prices may be little more than notional breakdowns of the contractor's tender offer and may have more of a marketing than a cost basis.

Practical building contractors are often scornful of this approach, and it is often suggested by people from the construction side that the cost planner should be more concerned with 'costs' (or 'real costs' as the proponents of this argument like to call them). This is an important issue and one which we are now going to look at, but it is not the simple choice between fiction and reality that is implied in this argument. In fact any figure in this context is simultaneously both a 'price' and a 'cost', it just depends where you are looking at it from. It can generally be said that 'the seller's price is the buyer's cost'. Thus:

- The contractor's price is the client's cost.
- The sub-contractor's price is the contractor's cost.
- The materials supplier's price is the sub-contractor's cost of materials.

Today, because so much of the work on major projects is being undertaken by sub-contractors on a price basis, and the independent professional quantity surveyor is increasingly involved in the management of projects, it is doubtful whether many general contractors now know much more about production costs than the quantity surveyor. But in a market-based economy the whole notion that there is such a thing as 'real costs' is a mistaken one anyhow.

For cost-planning purposes both the finished product price approach and the resource cost approach have their strengths and weaknesses, and good cost planners should understand these and know when each should be used, rather than simply adopting the one normally used by their profession in the past.

We have already looked at finished product prices in some detail, and now we need to understand how resource costs are in fact incurred.

The contractor's own costs

Builders like to think that they know about costs, and they think of them in resource categories, which are rarely thought of separately by traditional quantity surveyors as they manipulate their all-in unit rates.

Contractor's direct costs

Direct site labour

These are costs relating to the tradesmen and labourers actually producing the work. At one time these costs would have been described as 'wages', being almost entirely composed of this single item, but today they will include substantial payments in respect of National Insurance schemes, holiday schemes, training schemes, etc.

In the strictest sense of the word there may be almost no 'wages' paid to production employees at all, because so much of the work is done by sub-contractors. In addition, because for various reasons to do with taxation, and perhaps the inherent British dislike of the master/servant relationship, the custom grew up whereby the contractor's few production workers were 'self-employed' and were taken on as 'labour-only sub-contractors' for a fee. However, recent government legislation has inhibited this practice.

Labour costs are of particular concern to the contractor because they have to be paid out weekly in ready cash as they are incurred and cannot be postponed or put on a credit basis as can most other commitments – we will see later on in this chapter how important this is. It is convenient to include the 'labour-only' sub-contractors under this heading, as they will also usually require weekly cash.

Materials

The amounts comprising these costs will usually be paid by means of monthly credit accounts, payment being due at the end of the month following that in which the materials are delivered (so that materials delivered in January are paid for at the end of February – rather like credit cards).

Such settlement by the contractor usually entitles the firm to a 'cash discount', $2\frac{1}{2}$% being the most common figure although 5% may occasionally be allowed. It is not unusual for contractors to delay payment beyond the 'cash' settlement date, sometimes for a further month or even 2 months. Since they may lose their cash discount for the sake of 1 or 2 months' credit this could be an expensive way of borrowing money compared to a bank overdraft, but it has the advantage of being ready and convenient and of not requiring collateral security.

Within reason, and being careful not to let things get out of hand, builders' merchants are used to acting as financiers to the industry in this way (particularly in hard times). This has advantages for the merchant. If a building firm owes a substantial amount it is difficult for it to withdraw its custom from the supplier, or even reduce its level of buying substantially, as immediate settlement of accounts might be called for. The merchant therefore has a captive customer who cannot afford to be too fussy over prices or delivery dates.

In many cases the contractor will have been paid by the client for the materials, or for the work in which they are incorporated, before settling with his merchants. Because of the increasing tendency to buy in fabricated components rather than making things on site, a larger proportion of expenditure now falls into this category, and a smaller proportion into labour, than was previously the case. (Note that since most materials suppliers are not parties to the contract they are not entitled as regards payment to the protection of the 1996 Act mentioned on page 244.)

Small plant

This item covers hand tools and small mechanical plant of a kind that can be directly associated with specific pieces of finished work.

Sub-contractors and major specialist suppliers

Because of the increasing tendency to employ specialist sub-contractors rather than using the builder's own labour and materials, this category has expanded over the last 20 years and is now usually the largest head of expenditure. In terms of payment methods it very much resembles the 'materials' category, but sub-contractors have always tended to be less accommodating than materials suppliers with regard to extended credit.

If sub-contractors and suppliers are 'nominated' or 'named' by the architect they are in a position to complain to the architect if payment is delayed, and since such complaints cast doubts upon the builder's solvency, the builder will usually make efforts to pay them fairly promptly.

Again, the builder will often have been paid by the client for the work before settling with his sub-contractors, and some large firms of builders have attempted to formalise this in the past by inserting a 'pay-when-paid' clause in their sub-contracts. This practice has now been outlawed by the Housing Grants, Construction and Regeneration Act 1996 which, subject to certain exceptions, gives all the parties to a building contract the right to know the amount to be paid and the right to be paid on a determinable date.

Site indirect costs ('preliminaries')

The term 'preliminaries' is sometimes used to describe these items because they usually appear at the beginning of a BQ under this heading. They comprise all the items of site expenditure which cannot be attributed to individual items of work but to the project as a whole, or to substantial sections of it. Such costs may include:

- *Salaries and wages of site staff.* The salaries paid to management, supervisory and clerical staff employed on the site. In former times these payments would have been limited to a foreman and a few junior site staff, but today the total costs of on-site management will often be greater than those of the directly employed production workers on the site.
- *Site offices, messrooms and facilities.* Again, at one time these were a comparatively minor item, but today the temporary site office buildings can form a major multi-storey complex.

Major plant

Large items of mechanical plant may be dealt with in one of two ways:

- They may be charged to the job when brought on to the site and credited (less depreciation) when removed.
- Alternatively, an hourly or weekly hire charge will be made plus a charge for bringing the plant on site and removing it.

Most large contractors find it convenient to set up a subsidiary plant company, which will charge the plant out to the sites on a hire basis, while plant hired in from outside will be similarly dealt with and paid for by credit account. Lorry (truck) transport is usually arranged in a like manner.

It is unusual for a builder to own large items of plant except through a subsidiary company. There are substantial advantages in hire purchase through a finance company, since the interest charges (less tax) will be lower than the return which the contractor expects to make on the working capital which he employs; it would therefore be uneconomic for him to invest any of this capital in plant purchase.

It is difficult to attribute the costs of major plant items to specific pieces of direct work, and they are usually treated as a site indirect cost. However, they can sometimes be allocated to major cost centres, such as 'excavation' or 'concrete superstructure'.

Off-site costs (also called 'establishment charges' or 'overheads')

These represent costs incurred in running the company as a whole, and which cannot be attributed to any one particular contract – head office expenses, builder's yard, salaries of central management and directors, insurances and interest on loans.

Off-site costs are usually allocated to projects as a percentage of the direct costs. This can operate unfairly against small simple projects which require little head-office input, and some major contractors have separate 'small works' departments run on more economical lines, to avoid loading their smaller jobs with a large overhead and therefore making it difficult for them to compete with smaller firms.

Profit (also called 'mark-up')

Strictly speaking this is not a cost, it is in fact the difference between the builder's cost and the client's price.

Two typical examples

The following cost breakdowns relating to two large office buildings in the Home Counties of England in the mid 1980s, undertaken under ordinary lump-sum contracts by a main contractor, may be of interest. The figures for project B overstate the proportion of builder's direct work, since it was not possible to separate out some of the small sub-contractors undertaking traditional trade work.

Profit is included under the various cost heads, since because of commercial secrecy it was not possible to separate them out.

		Project A	Project B
(1)	Builder's direct costs	21.0%	37.0%
(2)	Sub-contractors and major specialist suppliers	70.0%	49.5%
	Total direct costs	91.0%	86.5%

(3)	Site indirect costs	6.5%	10.0%
	Total site costs	97.5%	96.5%
(4)	Off-site costs	2.5%	3.5%
	Total costs	100.0%	100.0%

Cash flow and the building contractor

The pattern of cash flow on a project is vital to a contractor. As an example a simple contract for a £410,000 building, to be erected in 30 weeks, has been chosen. The prime cost to the contractor is £400,000, leaving £10,000 profit. This represents $2\frac{1}{2}$% on turnover, which is fairly usual, but the real profit percentage may be very different to this miserable figure which is so often quoted as an example of the poor financial returns of the building industry.

Table 8 shows the weekly outlay on labour, plant hire and overheads, the weekly value of material deliveries, and the sub-contractors' accounts which are received monthly. The table also shows the total prime cost at the end of each month together with the quantity surveyor's valuation.

In this particular example it is assumed that 10% of the value of the work is withheld by the client each month as 'retention' until the total of the retention fund amounts to £25,000, after which remaining work is paid for in full. This arrangement would be less generous to the contractor than the current requirements of the British JCT Standard Form of Building Contract, and if this Form was being used the contractor would have rather more cash in hand than is shown in the examples.

It will be seen from Table 8 that, as usual, the job gets into its stride slowly, following the normal 'S curve' of expenditure plotted against time (Fig. 17.2), and because much of the early expenditure is in any case related to setting up rather than producing finished work, the first valuation by the quantity surveyor does not meet the full cost. However, by the time the job is halfway through it is showing a handsome profit (perhaps the 'loading' of BQ rates for the early trades may have helped) although the finishing off, again as usual, is not very profitable.

Table 9 illustrates the contractor's cash flow, on the assumption that everything goes as it should. The client pays the amounts of the valuations about a fortnight after they are made, and the contractor pays the nominated sub-contractors as soon as he receives the money. Materials are paid for at the end of the month following delivery (in practice the sub-contractors and materials suppliers would allow discounts for such prompt payment, but these have been ignored in the example in order to keep it reasonably simple). Labour, etc., has to be paid for weekly as the costs are incurred. At the conclusion of the project the client pays half of the retention sum and pays the balance 6 months later.

The cash flow in Table 9 is also shown graphically in Fig. 17.3. What is the first thing we notice about the cash flow as shown in Table 9? Although this is a £410,000 contract, and although the contractor is certainly bearing the risk (and undertaking the organisation) of a project of this size, such a figure has nothing to do with the contractor's financing of the job.

Table 8   Weekly outlays for a 30-week, £410,000 building.

	Week No.	Wages, plant hire and overheads	Materials delivered	Sub-contractors accounts received	Total prime cost and overheads	QS valuation	Valuation less retention
March	1	1,000	2,000				
	2	1,500	1,000				
	3	1,500	1,000				
	4	2,000	3,000		13,000	10,000	9,000
April	5	3,000	10,000				
	6	3,000	10,000				
	7	3,000	6,000				
	8	4,000	6,000	10,000	68,000	70,000	63,000
May	9	4,000	10,000				
	10	4,000	3,000				
	11	5,000	20,000				
	12	5,000	10,000				
	13	5,000	10,000	10,000	154,000	170,000	153,000
June	14	5,000	17,000				
	15	6,000	15,000				
	16	6,000	10,000				
	17	5,000	5,000	10,000	233,000	250,000	225,000*
July	18	5,000	5,000				
	19	5,000	10,000				
	20	4,000	5,000				
	21	3,000	3,000	30,000	303,000	325,000	300,000
Aug	22	3,000	2,000				
	23	3,000	—				
	24	2,000	5,000				
	25	2,000	—				
	26	2,000	5,000	30,000	357,000	375,000	350,000
Sept	27	4,000	10,000				
	28	4,000	—				
	29	3,000	—				
	30	2,000	—	20,000	400,000	410,000	385,000
	Total	106,000	184,000	110,000	400,000	410,000	385,000
						retention	25,000
							£410,000

*Maximum retention now withheld.

Except for 1 or 2 weeks the contractor never has more than £25,000 sunk in the contract and in fact often has no money invested in it at all – for example, by the middle of June the firm has received £43,000 more than it has paid out. If we therefore say that the firm could carry out the job on a working capital of £25,000, its £10,000 profit is not $2\frac{1}{2}$% but 40%, a very different figure!

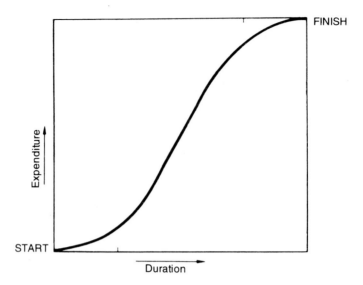

Fig. 17.2   Typical S-curve of project expenditure plotted against duration.

In fact the true situation is even better than this, because the contractor's average working capital (setting positive cash flows in some weeks against negative in others) is much less than £25,000, as can be seen from Fig. 17.3. This is therefore not a £410,000 job, as far as the contractor's budgeting is concerned.

Before we become too excited at the prospect of making 100% (or more) per annum profit as a matter of course, we should look at Table 10 and the accompanying Fig. 17.4. This represents the cash flow on the same project where the client is not being quite so helpful. The monthly payments are being delayed for a further 2 weeks, and the QS is being 'prudent' with the valuations, finally undervaluing to the extent of £10,000 (possibly because of variations which have not been agreed), although the full amount is eventually paid over.

This is still the same project with the same costs and the same profit, but as far as the contractor is concerned it is on a completely different scale. The contractor now has an almost permanently negative cash flow, often amounting to £30,000 to £40,000. This project will involve two or three times the capital commitment of the previous one and as far as the contractor's financing is concerned it is a project of more than double the size, although the profit is still the same. On the other hand, if the contractor were to complain the client might wonder what all the fuss was about. As Fig. 17.5 shows, the effect of the different payment pattern on the client's cash flow is proportionately very small.

Many contractors would confirm that this kind of situation is neither unusual nor does it by any means represent the worst that may befall them as regards deferring of payments, although the new Construction Act 1996 does now afford them some protection.

The contracting firm's remedy has often been to defer payment in turn to its own suppliers, which in the above example could more than restore the cash flow figures to their former satisfactory state (Table 11 and Fig. 17.6). It is possible

Table 9 Payments and receipts in £s based on Table 8.

	Week No.	Wages etc.	Materials	Sub-contractors	Total	Amounts received	Cumulative cash flow
				Payments			
March	1	1,000			1,000		−1,000
	2	1,500			1,500		−2,500
	3	1,500			1,500		−4,000
	4	2,000			2,000		−6,000
April	5	3,000			3,000		−9,000
	6	3,000			3,000	9,000	−3,000
	7	3,000			3,000		−6,000
	8	4,000	7,000 (March)		11,000		−17,000
May	9	4,000			4,000		−21,000
	10	4,000		9,000	13,000	54,000	+20,000
	11	5,000			5,000		+15,000
	12	5,000			5,000		+10,000
	13	5,000	32,000 (April)		37,000		−27,000
June	14	5,000			5,000		−32,000
	15	6,000		9,000	15,000	90,000	+43,000
	16	6,000			6,000		+37,000
	17	5,000	53,000 (May)		58,000		−21,000
July	18	5,000			5,000		−26,000
	19	5,000		9,000	14,000	72,000	+32,000
	20	4,000			4,000		+28,000
	21	3,000	47,000 (June)		50,000		−22,000
Aug	22	3,000			3,000		−25,000
	23	3,000		27,000	30,000	75,000	+20,000
	24	2,000			2,000		+18,000
	25	2,000			2,000		+16,000
	26	2,000	23,000 (July)		25,000		−9,000
Sept	27	4,000			4,000		−13,000
	28	4,000		27,000	31,000	50,000	+6,000
	29	3,000			3,000		+3,000
	30	2,000	12,000 (August)		14,000		−11,000
Oct	32			18,000	18,000	35,000	
		Release of retention		5,500	5,500	12,500	+13,000
	34		10,000 (September)		10,000		+3,000
April	60	Release of retention		5,500	5,500	12,500	+10,000
	Total	106,000	184,000	110,000	400,000	410,000	(10,000)

that delaying payment in this way might cause the contractor to forfeit some of the cash discounts which the materials suppliers give. These are normally of the order of 2½%. In the event of all the suppliers refusing to give any discount the contractor would stand to lose a total of £4,600, although in practice it would be unlikely that more than a few suppliers would do so unless the delays became very serious. However, the loss of a few cash discounts might be regarded as a

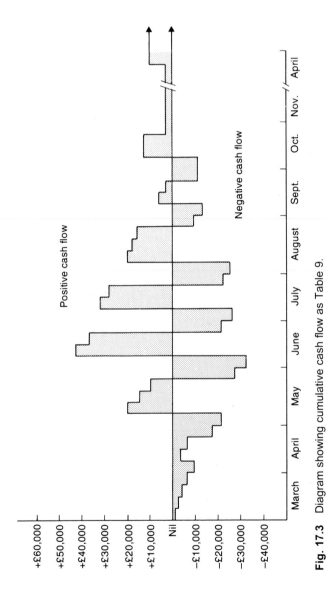

Fig. 17.3   Diagram showing cumulative cash flow as Table 9.

Table 10 As Table 9 but with late and underestimated payment in £s by client.

	Week No.	Payments				Amounts received	Cumulative cash flow
		Wages etc.	Materials	Sub-contractors	Total		
March	1	1,000			1,000		–1,000
	2	1,500			1,500		–2,500
	3	1,500			1,500		–4,000
	4	2,000			2,000		–6,000
April	5	3,000			3,000		–9,000
	6	3,000			3,000		–12,000
	7	3,000			3,000		–15,000
	8	4,000	7,000 (March)		11,000	8,000	–18,000
May	9	4,000			4,000		–22,000
	10	4,000			4,000		–26,000
	11	5,000			5,000		–31,000
	12	5,000		9,000	14,000	52,000	+7,000
	13	5,000	32,000 (April)		37,000		–30,000
June	14	5,000			5,000		–35,000
	15	6,000			6,000		–41,000
	16	6,000			6,000		–47,000
	17	5,000	53,000 (May)	9,000	67,000	88,000	–26,000
July	18	5,000			5,000		–31,000
	19	5,000			5,000		–36,000
	20	4,000			4,000		–40,000
	21	3,000	47,000 (June)	9,000	59,000	70,000	–29,000
Aug	22	3,000			3,000		–32,000
	23	3,000			3,000		–35,000
	24	2,000			2,000		–37,000
	25	2,000		27,000	29,000	75,000	+9,000
	26	2,000	23,000 (July)				–16,000
Sept	27	4,000			4,000		–20,000
	28	4,000			4,000		–24,000
	29	3,000			3,000		–27,000
	30	2,000	12,000 (Aug)	27,000	41,000	48,000	–20,000
Oct	34		10,000 (Sept)	18,000	28,000	34,000	
		Release of retention		5,500	5,500	12,500	–7,000
May	64	Release of retention		5,500	5,500	22,500	+10,000
	Total	106,000	184,000	110,000	400,000	410,000	(10,000)

small price to pay for the benefit of a cashflow so much improved that the project could run without any capital commitment at all after the first 16 weeks.

Finally, however, a dreadful warning: suppose we have an exactly similar job, where the client pays punctually and fully (as in the first example) but where the contractor has underpriced the work by 10% and is therefore not going to make

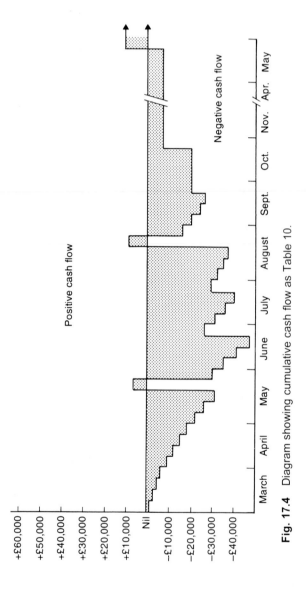

Fig. 17.4 Diagram showing cumulative cash flow as Table 10.

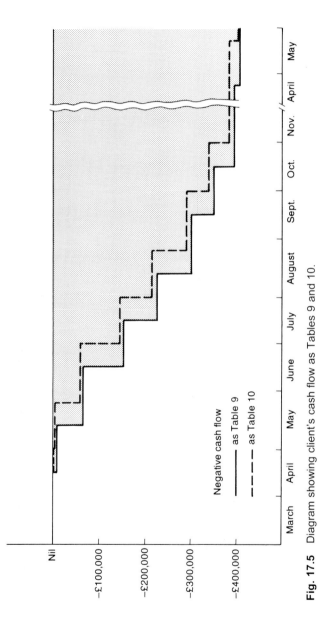

Fig. 17.5 Diagram showing client's cash flow as Tables 9 and 10.

Table 11   As Table 10 but with one month's delay in payments (in £s) to materials suppliers.

	Week No.	Wages etc.	Materials	Sub-contractors	Total	Amounts received	Cumulative cash flow
			Payments				
March	1	1,000			1,000		−1,000
	2	1,500			1,500		−2,500
	3	1,500			1,500		−4,000
	4	2,000			2,000		−6,000
April	5	3,000			3,000		−9,000
	6	3,000			3,000		−12,000
	7	3,000			3,000		−15,000
	8	4,000			4,000	8,000	−11,000
May	9	4,000			4,000		−15,000
	10	4,000			4,000		−19,000
	11	5,000			5,000		−24,000
	12	5,000		9,000	14,000	52,000	+14,000
	13	5,000	7,000 (March)		12,000		+2,000
June	14	5,000			5,000		−3,000
	15	6,000			6,000		−9,000
	16	6,000			6,000		−15,000
	17	5,000	32,000 (April)	9,000	46,000	88,000	+27,000
July	18	5,000			5,000		+22,000
	19	5,000			5,000		+17,000
	20	4,000			4,000		+13,000
	21	3,000	53,000 (May)	9,000	65,000	70,000	+18,000
Aug	22	3,000			3,000		+15,000
	23	3,000			3,000		+12,000
	24	2,000			2,000		+10,000
	25	2,000		27,000	29,000	75,000	+56,000
	26	2,000	47,000 (June)		49,000		+7,000
Sept	27	4,000			4,000		+3,000
	28	4,000			4,000		−1,000
	29	3,000			3,000		−4,000
	30	2,000	23,000 (July)	27,000	52,000	48,000	−8,000
Oct	34		12,000 (Aug)	18,000	30,000	34,000	
		Release of retention		5,500	5,500	12,500	+3,000
Nov	38		10,000 (Sept)		10,000		−7,000
May	64	Release of retention		5,500	5,500	22,500	+10,000
	Total	106,000	184,000	110,000	400,000	410,000	(10,000)

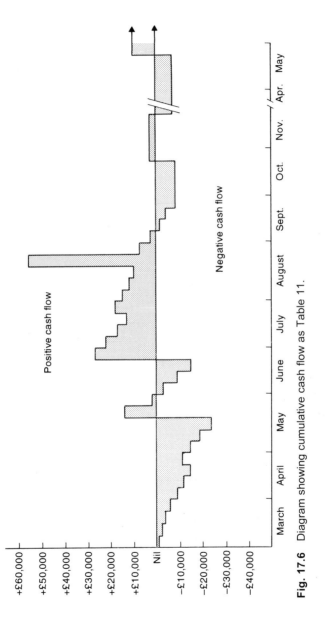

Fig. 17.6 Diagram showing cumulative cash flow as Table 11.

£10,000 profit but £20,000 loss. Also, because of financial difficulties, the contractor is paying the suppliers one month late. The resultant cash flow (Table 12 and Fig. 17.7) looks very satisfactory until the end of the project in September. It is in many ways better than that of the soundly managed contractor in Figs 17.3 and 17.6, and requires a working capital of about £10,000 for a few weeks only. However, instead of a handsome profit this contractor will finish by losing the whole of his working capital twice over.

If the firm has several such jobs going on concurrently, so that the negative cash flows on one coincide with the positive flows on another, it will be able to keep trading for some time before the crash comes, and meanwhile its cash flow figures may look fairly healthy.

It will be seen how difficult it is to distinguish the early symptoms of insolvency when looking at a contractor's accounts, and indeed it is quite possible that a badly managed contracting firm may not itself realise what is happening until it is too late.

When times are bad in the construction industry it is common for contractors to submit very low tenders in order to obtain work and so keep their organisation employed, and they can then easily become short of cash. It is also not unknown for contractors who have already got into this position, either for the above reason or through bad management, to continue to quote absurdly low prices. This is because they urgently need the cash flow from new work in order to pay their past debts.

If the contracting firm in the above example is receiving payments on another job by September it will be able to pay its suppliers and keep going (although the new job in turn will be getting into even worse difficulties in due course and will need an even more drastic dose of the same medicine).

Quantity surveyors do well to be suspicious of a building company which is expanding its operations rapidly at prices which its competitors cannot match. However, we can see from the earlier examples that even a soundly managed contracting firm will find it very tempting to get deeply into debt at the bank if it is able to get returns of more than 50% on money which it is borrowing at less than 15%.

So cash flow assessment of accounts must be used carefully, but these examples do demonstrate that quite small percentage differences in estimating can have a dramatic effect on profitability. This is in fact the principal weakness of the construction industry, that the difference between a substantial profit on capital and a substantial loss can lie inside the normal margin of error in estimating.

A greater capital investment in a project on the part of contractors might lead to greater stability, as the required profit on turnover would then have to be much higher than 1% or 2%. If the contractor had to make 10% profit on turnover to get a reasonable return on capital, then estimating errors of 2% or so would have a proportionately smaller effect on the firm's overall profit percentage and would not make the difference between 'boom or bust'. Moves to reduce 'retentions' in recent times have had exactly the opposite effect. However, retention bonds are now coming into vogue.

Table 12 As Table 9 but builder's work underestimated by 10%, and one month's delay in payments to materials suppliers. Values in £s.

| | Week No. | Payments | | | | Amounts received | Cumulative cash flow |
		Wages etc.	Materials	Sub-contractors	Total		
March	1	1,000			1,000		−1,000
	2	1,500			1,500		−2,500
	3	1,500			1,500		−4,000
	4	2,000			2,000		−6,000
April	5	3,000			3,000		−9,000
	6	3,000			3,000	8,100	−3,900
	7	3,000			3,000		−6,900
	8	4,000			4,000		−10,900
May	9	4,000			4,000		−14,900
	10	4,000		9,000	13,000	49,500	+21,600
	11	5,000			5,000		+16,600
	12	5,000			5,000		+11,600
	13	5,000	7,000 (March)		12,000		−400
June	14	5,000			5,000		−5,400
	15	6,000		9,000	15,000	81,900	+61,500
	16	6,000			6,000		+55,500
	17	5,000	32,000 (April)		37,000		+18,500
July	18	5,000			5,000		+13,500
	19	5,000		9,000	14,000	65,700	+65,200
	20	4,000			4,000		+61,200
	21	3,000	53,000 (May)		56,000		+5,200
Aug	22	3,000			3,000		+2,200
	23	3,000		27,000	3,000	70,500	+42,700
	24	2,000			2,000		+40,700
	25	2,000			2,000		+38,700
	26	2,000	47,000 (June)		49,000		−10,300
Sept	27	4,000			4,000		−14,300
	28	4,000		27,000	31,000	48,000	+2,700
	29	3,000			3,000		−300
	30	2,000	23,000 (July)		25,000		−25,300
Oct	32			18,000	18,000	33,500	
		Release of retention					−3,900
	34		12,000 (Aug)		12,000		−15,900
Nov	38		10,000 (Sept)		10,000		−25,900
April	60	Release of retention		5,500	5,500	11,400	−20,000
	Total	106,000	184,000	110,000	400,000	380,000	(−20,000)

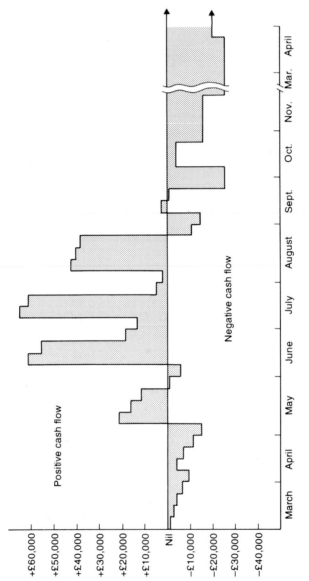

Fig. 17.7 Diagram showing cumulative cash flow as Table 12.

Allocation of resource costs to building work

The contractor would appear to be in a much better position than the consulting quantity surveyor as far as knowledge of actual building costs is concerned, but this is not necessarily the case. Keeping an accurate record of costs in the four categories:

- builder's direct costs (site labour, materials, small plant),
- sub-contractors and major specialist suppliers,
- site indirect costs and
- major plant

is not very difficult, but translating them into usable data for cost planning future projects is another matter.

The contractor would in fact be ill advised to use the actual costs from a particular project for cost planning a quite different one. There are two reasons for this. First, many of the factors which affect site costs on an individual project have nothing to do with the design of the building and will not repeat from job to job. These include:

- weather;
- supervision;
- industrial and personal relations;
- obstruction by other trades;
- the skill with which the work is planned and organised;
- alternatively, lack of clear instruction;
- waiting for delivery of materials;
- accidents;
- replacement of defective work;
- failure by sub-contractors;
- psychological pressures.

Secondly, in practice, costs are rarely kept in any greater detail than the 'activity' or 'operation'. An operation has been described by the Building Research Establishment as 'a piece of work which can be completed by one man, or a gang of men, without interruption by others', such as the whole of the brickwork to one floor level, or the whole of the first fixings of joinery. Unfortunately, each operation is unique to the project, and the cost information cannot be re-used in this form.

System building, prefabrication and cost

No discussion of building costs would be complete if it did not deal with the effect of system building and prefabrication on costs. This was widely promoted in the 1960s as the answer to many of the industry's problems, but a number of unfortunate experiences caused the inevitable reaction against it.

System building involves varying degrees of discipline. In its most simple form it may only be a standardised form of more-or-less traditional building, saving time and money by the familiarity of the standard construction and detailing, the repetitive use of formwork units, standard components, and so on.

It is difficult to obtain competitive tenders for this sort of work because, if prices are given by two or more firms, each firm will quote for its own patent system and it will be almost impossible to compare the value of the tenders without a considerable amount of work. In these circumstances cost comparisons based on cost modelling techniques of the type described in this book may provide a truer picture than an attempt to check detailed quantities and prices, even where these are provided.

However, in its more highly developed forms, system building may involve the delivery of the whole building in the form of a kit of parts ready to fit together in a few days (in the case of a house) or in a few weeks in the case of a larger building.

The economics of this practice are not easy to evaluate, partly because of the advertising and hard selling which often go with it, and partly because the economics are in fact much more complicated than the more usual site building costs.

The cost advantages which the prefabricating firm possesses are, at first sight, considerable:

- It has the benefit of planned mass production under factory conditions, safe from the weather hold-ups which affect site output and free from the difficulties of supervision and quality control which occur when operatives are working all over a scattered site.
- It can employ expensive but money-saving plant which would be too cumbersome, valuable or delicate for site use, and which in any case could not find full employment for its output on a single site.
- It can take advantage of modern methods of handling and transportation to bring its large fabricated units direct from the factory to the position on site where they are required.

In fact we might tend to accept the view that the lack of development of factory-based building techniques simply shows that the construction industry is hopelessly old-fashioned.

Economic problems of prefabrication

However, there are very sound reasons why prefabrication has had a much smaller impact upon the industry than was at one time expected. One reason has been the popular dislike of an environment composed of factory-produced buildings, although this might have been easier to overcome if there really had been a strong economic justification for them.

In fact the economic gains have usually proved to be disappointing. What are the reasons for this?

Traditional factors in favour of site production

These were always known about, and it was hoped that they would be more than offset by the above advantages.

- The site builder has none of the heavy overheads of factory production (the firm will not have to pay rent for its site workshops and will probably not even have to pay rates).
- Many of the cheap techniques (such as bricks-and-mortar) which the site builder uses are not suitable for off-site production.
- The crude handling methods which can be used with the small and rough components used for site assembly are cheaper than the tackle required for the careful handling of large units.
- The site builder is able to provide a 'made-to-measure' building instead of one 'off-the-peg'.

Other economic problems which have come to light with experience of prefabrication

- A broken brick is simply a broken brick and there are plenty of unbroken ones to use, but if the corner of some special component is damaged another one will have to be ordered.

The cost of replacement is the least part of the difficulty; the disruption to programme caused by the resultant delay can be far more serious.

- The disadvantage caused by the fact that only the building superstructure can be prefabricated and the infrastructure has to be produced on site is much greater than was supposed.

On low-rise projects such as housing, factories, health buildings, etc., the site preparation, levelling, foundations, drainage and site services, roads, paths, car parks, fencing, landscaping, etc., involve so much work on site and site organisation that the prefabrication of the basic superstructure is only dealing with part of the problem and not always the most significant part. This particularly applies if the finishings and services are not even included in the package.

This problem is tending to increase rather than decline in importance, because today there are very few projects where the site is a level open field on which a factory-produced building can easily be placed – in fact the arrangement of the building or buildings is more usually dictated by the shape and configuration of the site.

The building of the various in-situ connections between standard units (especially if these are at different levels on an undulating site), or the work necessary to accommodate them to irregular site boundaries, is piecemeal work which is difficult to organise efficiently, and it can more than swallow any cost savings generated by factory production of the main units. In such circumstances it might well have been better to build an in-situ building designed from the start to fit its location.

An attempt to design prefabricated bathrooms for houses and flats in the 1960s failed largely for this reason – when all the other partitions in the dwelling had to be built and finished on site, and all the other doors had to be hung there was not really any advantage in having had two partitions and one door fixed in the factory, and the work of handling this large unit, and fitting it into the in-situ construction, was simply extra cost. The people putting forward this scheme just had not thought it through.

Prefabricated 'toilet pods' have recently come back into vogue for commercial buildings and other large projects, but this situation is quite different. Where most of the floor area is simply open space for letting, the toilets are on quite a different scale to the rest of the building with their plumbing and generally more domestic human scale. In addition, the programming advantages of having such work done off-site while the rather different site work is in progress are considerable.

- A more important and less obvious handicap which faces the prefabricating firm is the large capital investment which has to be made in developing and testing the system, establishing and equipping the factory and, most of all, stockpiling the components until they are bought by the customer.

There is little opportunity to obtain the very high rate of return on capital which the site builder is able to obtain on his much more limited capital investment (as explained above), and contractors have not found this to be a worthwhile field for deploying their financial resources.

- Even where it is possible to produce a scheme for prefabrication which shows at any rate a reasonable return on capital, there are still two further difficulties to contend with.

The first is the actuarial risk of some major snag developing in the system once it is tried out in the harsh conditions of the real world. This could involve both the costs of putting things right and the possible loss of custom resulting from bad publicity; it is not really practicable to make adequate allowance for this risk in a pricing structure.

The second difficulty is the one which has been most often quoted by firms who have been unsuccessful in this field. Factory-produced buildings need a large and assured market in order to be competitive in cost, as do all factory products which involve a heavy initial expenditure on plant, tooling and design, and preferably demand should not fluctuate too wildly year by year.

These conditions are not met by the construction sector which is notoriously subject to boom and slump as a result of either economic forces or, more often, government fiscal policy.

The flexible site-organised building industry has enough trouble as a result of 'stop-go', but it is absolutely fatal to the capital-intensive factory production of buildings. Those firms from outside the building industry who moved into this field were soon forced to abandon it.

It is interesting that the motor industry, again held up as an example to the construction industry in the Egan Report of 1998, has nevertheless wisely refused at any time in the last 40 years to get involved in factory production of buildings.

Suitable and unsuitable fields for prefabrication

The only sector where prefabrication of complete buildings seems to have established the right conditions for its survival is in the field of temporary and light industrial buildings (sheds, garages, temporary classrooms and offices, church halls, single-storey factories and warehouses, etc.) where the pre-fabricated building dominates the market.

However, in many ways the production, distribution and erection of these light units does not involve the same difficulties that occur with larger and heavier components, but it should be noted that the firms in this sector are tending to use their experience to progress to larger and more complex structures. Prefabricated student accommodation blocks are a recent example.

A piece of false reasoning

An erroneous justification for prefabrication is that the industry may be unable to cope with the volume of work required at busy times because of a shortage of labour, particularly qualified tradesmen, to work on site. This reason was in fact used to try and justify some of the prefabrication disasters of the 1960s and early 1970s – 'We had no alternative'.

As we have just seen, however, prefabrication cannot be the answer to short-term 'boom' conditions of this nature which are exactly the opposite of the settled market which it needs for survival. It was thought at one time that prefabrication might be forced on society simply because people were going to prefer working in the sheltered and well organised conditions of a factory to working on a building site in all weathers. However this is turning out not to be the case and it would be difficult to reproduce the currently popular site self-employment pattern of work in a factory.

A cool look

Prefabrication has gone through the stages of fashionable popularity and the ensuing disillusionment, so it is now perhaps possible to look more dis-passionately at its advantages and disadvantages, and to sort out the problems which are inherent in prefabrication from those which have simply resulted from poor implementation.

Today prefabrication is in fact widely used, although not usually in the context of a complete building system apart from the exceptions already mentioned. The structure of a modern building is usually produced on site, although it could be argued that the traditional steel frame is an example of prefabrication – and a particularly interesting one since the actual components are cut and finished to individual design requirements at the factory from standard sections.

However, it is in the fitting out of the building that prefabrication is more usually encountered and the tendency is for its influence to increase. Today windows are not merely fabricated off-site complete with fittings, but are often delivered preglazed and decorated. Similarly doors often come already hung in their frames and fitted with hardware, and although wrapped in protective material these components generally require more careful treatment than is given to general construction materials. Also, of course, the hydraulic and engineering services largely comprise finished components joined together by piping or cabling on site.

As already mentioned, there has even been a return to prefabricated toilet modules, although in the context of high-grade office buildings rather than housing. The reasons for this return of prefabrication include the ability to obtain a high standard of finish using factory processes, and especially the ability to speed up the work programme because the manufacture of these units can proceed in parallel with the structural construction of the building. These units are usually purpose-made and are probably more expensive than site-produced work – the saving comes from the telescoping of the construction period, and this may well be the sole advantage of prefabrication in many cases.

In any comparison between different forms of patent construction, or between factory and site production, it is also necessary to consider economic life and maintenance costs as well as first costs, and this will be even more difficult than usual because these factors will be pure guesswork as far as non-traditional materials or finishes are concerned.

The prudent architect or cost planner will need to be a good deal less optimistic than the salesperson when assessing the probabilities. The very expensive maintenance, and sometimes premature demolition, which proved necessary with some systems because of difficulties at the joints between components which were quite unforeseen when the systems were evaluated.

Summing-up

The construction industry is exceptionally flexible in dealing with local and national booms and slumps, but it adjusts its prices accordingly. Contractors' costs and prices therefore do not move in step with each other, except in very settled times – and it is the contractor's price that is the client's cost.

Contractors' costs comprise:

- direct costs (labour and materials);
- sub-contractors and major specialist suppliers;
- site indirect costs ('preliminaries');
- major plant;
- off-site costs ('establishment charges' or 'overheads').

The building industry works on a very small capital commitment compared to its turnover, and its cash flow pattern is therefore vitally important. Prefabrication of complete buildings (except single-storey warehouses, sheds and the like) has not so far proved to be of economic benefit.

Further reading

Harvey, R. & Ashworth, A. (1997) *The Construction Industry of Great Britain*, 2nd edn. Butterworth Heinemann, Oxford.
The Stationery Office (published annually) *General Economic Data. Britain*. The Stationery Office, London.

Chapter 18
Resource-Based Cost Models

Effect of job organisation on costs

Many major costs on building projects are not directly related to the quantity of work produced but are concerned with time and with occurrences (or non-occurrences) of various kinds.

The main quantity-related cost is materials, so that given prudent buying and no unreasonable wastage there should be little variation in the cost of this part of the work whichever contractor is appointed and however the work is organised.

The real scope for saving (or wasting) money lies in the non-quantity-related items, and these depend on the way the job is organised and managed. This is where the real competition between contractors takes place, especially today when, as we have seen, the contractor's main stock-in-trade is management skill.

A well managed construction project

A well managed project is one where both plant and workers have clear uninterrupted flows of work.

Money is not spent on removing or returning workers or plant from the site, or on unproductive time waiting on site because of gaps in the work flow, or on moving workers or plant unnecessarily around the site. Materials are channelled to the spot where they are needed at the right time, and with a minimum of double handling.

It is not merely a question of the actual time spent hanging around – people work much more productively when they can see a clear task in front of them and where they feel they are participating in an efficient operation. It is a constant complaint by contractors that they are usually unable to recover the cost of disruption of this state of affairs when delays or interruptions are caused by the client or the design team.

Traditional vs resource-based methods of cost planning

Seeing the increasing importance of these management-based costs gives scope for comparing the cost planner's traditional price-oriented approach unfavourably with a resource-based method which can take account of method and work flow. But the traditionalist's reply must be that traditional estimates are based on successful past tenders, and that few successful tenderers will have worked out their prices on the basis that the job will be badly run, whatever may have happened afterwards! In fact it might be claimed that the traditional price-based method automatically allows for good management without ever having to bother about the details of its implementation, and this facility is especially valuable in the early stages of estimating before the project has been fully designed.

Value added tax (VAT)

Value added tax is currently payable on building work other than housing, and also on architects', quantity surveyors', engineers', planning supervisors' and other professional fees. The rate at the time of writing is 17.5%.

It is customary to exclude this amount from estimates and tenders. This practice is well understood within the construction industry, and has been followed in this book. But this must be clearly pointed out in any figures given to clients, who may otherwise think that the estimate represents their total liability.

Resource-based cost models

'Real costs'

The detailed cost models which we have examined so far have been largely based upon the measurement of finished work in place, and its valuation from BQs or other market-price-orientated data. Quite apart from the general fallibility of BQ rates, these models have embodied two major fallacies:

- that the production cost of a building element is proportional to its finished quantity;
- that the cost of a building element has an independent existence which can be considered separately from the rest of the building.

We have been aware of these drawbacks almost from the start and are able to come to terms with them through the exercise of professional skill. However, it is worthwhile considering whether it would not be better to base our estimates and cost control procedures on production cost criteria, as is done in most (perhaps all) other industries.

One major reason why we do not normally adopt this approach has already been given, which is that under lump sum contracting arrangements it is the market price, rather than production costs, which concerns clients and their advisers. This reason, however, is not valid if we are considering a cost-reimbursement or management type of contract, or if we are looking at a situation where the cost planning is being carried out within a design-and-build organisation.

There is, however, another and quite different ground on which resource-based cost planning can be criticised. As has already been pointed out, much of the resource cost of modern building is attributable to plant and organisation. Both the initial estimating and the subsequent refining of the estimate therefore require the envisaging of technological solutions. Whether the building will be:

- steel-framed;
- precast concrete frame and panel or
- in-situ concrete

will demand totally different approaches with probably significant cost differences for any particular configuration and set of user requirements.

This last paragraph in fact may sound like an argument in favour of a resource basis rather than a criticism, but the big disadvantage is that it moves the design considerations of a building into the production field much too early in the process.

An economical structural system may be postulated by the resource-based cost planner and the configuration of the building developed, for example, to suit the radius of action of the tower crane (or cranes) placed in the most efficient positions, or to suit the repetition of precast units.

The architects's traditional approach, as we have already seen, is the opposite one – the form of the building is evolved primarily as a set of user-oriented spaces, and the best technological solution for that configuration is then investigated.

Quite clearly, in both cases, some compromises may have to be reached, but the basic issue is whether

- user needs and environmental considerations come first and construction methodology follows, or
- production efficiency should be the consideration from which design develops.

There are too many existing buildings which remind us that construction optimisation lasts for months but the consequences last for years (although not always as many years as the designers intended). Unfortunately production managers are no less lazy than the rest of us and prefer to postulate easy solutions rather than applying themselves to devising an efficient way of building a user-oriented design.

It ought to be possible to take this latter approach provided the details of construction have not been developed too far before the builder is called in. However, while it is true that the quantity surveyor's traditional elemental cost planning approach enables the early design process to proceed with some semblance of cost control prior to construction decisions, such decisions do have to be faced sooner or later.

Once this point is reached a resource-oriented approach has much to recommend it in cases other than the competitive price-in-advance situation, particularly where the contractor who will be undertaking the work is a party to the cost planning exercise. If this is not the case, then the method is of dubious advantage; there are many different ways of organising a building site and each contractor has his preferred methods and equipment, so that there is unlikely to be a 'best' solution that any builder would automatically accept.

Much of the advantage of a resource-based estimate is that it can be used for production-cost control purposes, and this benefit will be lost if the estimate has been prepared by someone else and the chosen contractor throws it out of the window. A resource-based estimate will deal quite separately with the different cost components of:

- labour,
- plant,
- materials and
- sub-contractors

rather than amalgamating them into a series of 'all-in' rates as is the quantity surveyor's custom. However, having said this, a very substantial part of the total cost will comprise sub-contractors' work.

Much of the resource-based estimate will therefore have to be based on the specialists' all-in prices in exactly the same way as a product-based estimate. So the claim by the builder's estimator to know more about the so-called 'real cost' of building than the independent professional cost planner becomes somewhat questionable. The resource-based estimator's immediate concern will therefore be with that part of the work which is usually undertaken with the contractor's own resources, normally the structure of the building including excavation.

The resource-based estimator's real expertise (and the area where the product-oriented estimator is weak) is in determining and pricing the site organisation and the project duration, including the management of all the specialists' programmes and the integration of their work with the building structure and with each others' efforts.

In a design-and-build, contracting or construction management organisation the estimator will almost certainly have the assistance of a planning engineer or project manager in this work. They may look at alternative configurations for the proposed building, as well as alternative means of obtaining the same configuration.

As far as estimating the cost of the contractor's own work is concerned the materials present few problems, since unlike labour and plant their cost is reasonably related to the quantities of finished work. Anybody preparing a resource-based estimate will, therefore, have to start off by measuring (or assuming) quantities of finished work in order to determine material requirements and also as a first step in looking at the scale of the project and the distribution and inter-relation of its parts. The person doing this will tend to work in bulk quantities (e.g. cubic metres of concrete) split into categories and locations which seem to be organisationally significant.

Before proceeding much further however, an outline programme for the works will need to be prepared.

Resource programming techniques

The estimator, in respect of each different estimate, will need to decide the principal operations to be undertaken and their methodology and duration. The operations cannot be considered in isolation since they will be inter-related by two factors:

- The need to use labour and plant effectively, so that operatives and machines do not stand idle for long periods between tasks, are not required to be working in two different places at once and do not spend too much time moving from one part of the site to another.
- The inescapable sequence of building, so that, for example, the walls and columns cannot be built until the foundations are completed, and the first floor cannot be placed until the ground floor walls and columns have been built.

There are various techniques in common use to assist the estimator in this task, which tend to rely in the first instance upon graphic methods.

The bar chart (or Gantt chart)

The traditional method is the 'bar chart'. On this chart a horizontal time scale is used, often divided into weeks, and the various operations comprising the project are listed vertically down the left-hand side. The timing and duration of each operation is then indicated by a horizontal bar spanning the relevant period of weeks and shown on the same line as the operation it refers to. An example is shown in Fig. 18.1, and it will be seen that the bars follow the classic pattern of moving diagonally from the top left-hand to the bottom right-hand of the chart.

The bar chart is simple and easy to follow. It gives quite a good picture of the way in which the various operations fit into the total contract period and is very popular on building sites for the purpose of monitoring progress and forward ordering.

However, although it is a good communication tool it has serious drawbacks as a planning tool, because:

- It does not help in determining the duration of operations (it cannot be drawn until these have been decided).
- It does not bring the interdependence, or otherwise, of the operations to the notice of the planner.

It therefore tends to show the results of planning which has been undertaken by other means. At one time these other means were usually informal, but in present-day practice the network diagram is commonly used.

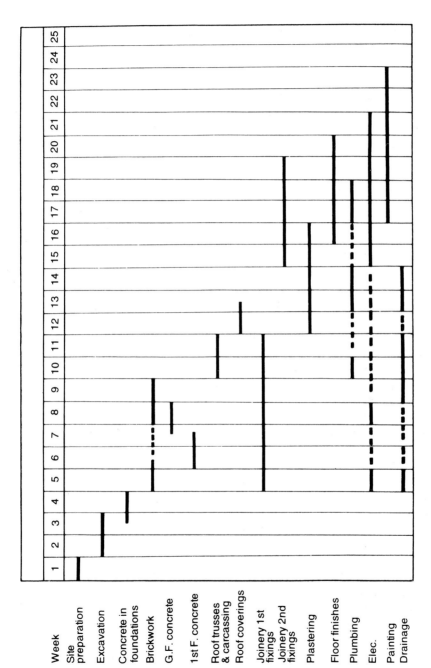

Fig. 18.1 A bar chart.

The network diagram

'Network diagram' is a generic term and also includes 'precedence diagrams' and 'critical path diagrams', which are special uses. The network consists of a series of nodes joined together by lines or arrows, and normally moves from a start at the left-hand side to a finish on the right.

In one system (Fig. 18.2), the lines or arrows represent operations (or activities) and the nodes represent the interdependencies which occur at their start and finish. A network prepared in this manner has the advantage that it may be drawn to scale with the lines or arrows of a length proportionate to their time requirements, and can develop into something that is almost as easy to read as a traditional bar chart.

However, it has drawbacks:

- Like the bar chart it cannot be drawn until the durations are known.
- Once drawn, it is very difficult to amend if durations change during the progress of the work.
- It can be difficult to follow, particularly as it is usually necessary to draw a number of 'dummy' arrows in order to close the network.
- There is always a tendency to see the length of each arrow as having some significance, even when this is not the case.

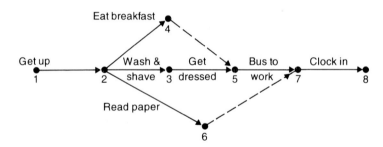

Fig. 18.2 Simple 'arrow = operation' network to illustrate going to work in the morning. Each node represents the beginning or end of an operation. Note the need for dummy (dashed) arrows. You cannot catch the bus to work until you have dressed and had breakfast. The paper need not be read until the bus ride.

For planning purposes, the preferred alternative approach (Fig. 18.3) is one in which the nodes represent the activities and the lines or arrows joining them simply illustrate dependencies. The nodes are usually drawn in the form of sequentially numbered circles which are large enough to contain some numeric information; the length of the lines joining them is of no significance and is chosen arbitrarily to suit a clear layout of the diagram.

The diagram can thus be drawn without any idea of the length of time which any operation will take and in this form reflects the earliest planning stage, in which the planner is identifying those operations which fundamentally depend upon each other and those which do not. Once this has been done it will never

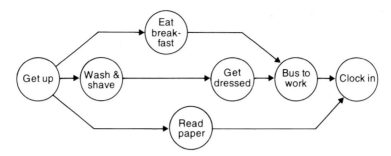

Fig. 18.3 The same project on the 'node = operation' system. The arrows indicate precedences. Note that in neither case does the diagram mean that breakfast is eaten whilst washing and dressing, simply that it is not essential for you to be dressed before eating – you might have your breakfast first. Similarly, the paper need not be read before you get on the bus, but you have the option of looking at it any time after you rise.

need to be redrawn (unless the project itself alters); any changes will be made only to the numeric information within the nodes.

A network in an unquantified form is simply a 'precedence diagram', showing which operations must precede which. No operation can take place until all operations arrowed into it have been completed.

Remember, on a building project, it is fundamental construction dependence which counts and not convenience or the efficient use of resources, which are considered at a second stage.

Two important points are worth noting:

- The network can be expanded to include not merely the construction but also the whole of the planning and design process and can thus become a tool in the hands of the client's professional representatives. This capability is particularly useful where overall time requirements are important or where design and construction are to run in parallel on a management, design-and-build or fast-track contract.
- The operations may be shown at a strategic level ('build walls') or at a much more detailed level ('build ground floor external walls', 'build in sills and lintels', and so on).

For eventual control purposes a fully detailed network may be used, but this is expensive to prepare and would be inappropriate at planning stage, before the actual detailed design has been finalised. On the other hand, a simple 'planning' network cannot show the interdependence of meshing operations. For instance, if the erection of precast beams can commence after only some of the columns have been completed this would require the work of each group (or even pair) of columns to be shown as separate operations.

This problem occurs throughout the project, but at planning stage is only likely to be considered in those parts which the planner feels to be of crucial importance – probably major structural work rather than finishes.

Once the network has been drawn and the precedences and dependencies settled it is time to quantify the problem. The procedures for doing this are sufficiently standardised for there to be a number of computer packages available, and the work then entails filling in the estimated durations on a schedule.

If manual methods are being used, however, a suitable method is to mark durations in the node circles which are divided into four quadrants for the purpose of quantification (see Fig. 18.4). In the *top left-hand* quadrant is written the reference number of the operation concerned (full description of the operation probably being written out in a numbered list) and in the *top right-hand* quadrant the estimated duration in days, weeks or whatever unit is being used. When all the individual durations have been inserted the total project time can be calculated.

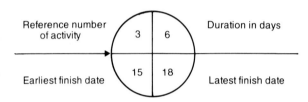

Reference number of activity   3 | 6   Duration in days

Earliest finish date   15 | 18   Latest finish date

Fig. 18.4   Activity node (showing division into quadrants for reference).

Beginning at the 'start' and working along the arrows the shortest possible elapsed time to the completion of each operation is inserted in the *bottom left-hand* quadrant. This is arrived at by adding the duration of the operation concerned to the highest elapsed time of any of the other operations which precede it and which are therefore arrowed into it. By the time 'end' is reached the total time for the project will have been calculated.

In respect of each operation, we will now know the earliest time at which it can be completed. Where the operation concerned is a crucial one this will also be the latest time at which it can be completed if the project as a whole is to be finished on the calculated date. However, many less important operations may be able to be delayed without delaying the completion of the whole project.

In order to identify the crucial operations the *bottom right-hand* quadrant of each node may be used to show the latest possible date at which the operation can be completed without causing delay. This is calculated by working backwards from the finish and, in the case of each operation, calculating from the earliest start time of any of the operations which immediately follow it.

When this task is completed it will be found that for some operations the earliest and latest dates for finishing are the same. These are the critical operations, and the sequence of arrows which joins them forms the so-called 'critical path' through the project.

In the case of a non-critical operation the difference between the earliest and latest finishing date represents the 'float', or the period for which that operation can be delayed without affecting the completion date for the project.

All these figures can of course be set out in the form of a schedule, instead of on the diagram itself. It should be noted that if part of the 'float' for a particular operation is actually utilised this may affect the float available for succeeding operations. If all the float is used up then the operation, and its successors, will have become critical and the critical path will have changed. It is therefore necessary for the project controller to keep a close eye throughout the progress of the job on operations with a very small float, as well as those which are already on the critical path.

Resource smoothing

There may well be two or more operations in a network which do not depend upon each other in the construction sequence – for example, foundation excavation and drainage trenches. It will not matter which is done first, and they could in theory be done simultaneously (whereas the foundation excavations and the concreting of them could not).

However, their relative timing may depend upon resource utilisation. To dig the foundation and drainage trenches simultaneously would involve bringing an unnecessary number of diggers to the site, whilst to leave several weeks between the two operations would involve either machines standing idle or else taking them away and bringing them back. If both operations were on critical or near-critical paths then the additional expenditure would have to be accepted (or the time for completion altered), but otherwise their floats would be adjusted to give a more economical sequential programme.

The adjustment of a programme in this way to make the best use of expensive plant, and to give gangs of operatives a consistent and balanced work pattern, is an important part of the planning exercise, and is often called 'resource smoothing'.

Resource-based techniques in relation to design costplanning

The use of resource-based methods is not practicable at the earliest budgeting and planning stages as the cost planner who is intending to use these methods needs to start off with some idea of the configuration of the proposed building – its size, shape, height and preferred technology.

The cost planner may well suggest a suitable combination of these requirements, and if working for a construction firm these may be based on systems already rationalised in the organisation and which can be built quickly and efficiently. For some types of development (e.g. industrial premises) this could be a valid approach. Alternatively the cost planner may begin with a solution proposed by the architect, test it in terms of resource use and then investigate alternatives.

At this stage the main concern will be with those things which can be identified as major time constraints and production cost constraints. By using skills derived from experience the cost planner will be looking for a solution which gives effective use of labour and equipment by providing smooth and continuous work flows, and which enables the most economical type of plant to be used.

For example, if one single heavy component has to be lifted at the extreme reach of the tower crane this will dictate a much heavier and more expensive type of crane to hire and run, probably for the whole duration of the contract. The crane 'cycle' will also dictate the intervals at which components, or skips of wet concrete, can be lifted into position, which in turn may determine the size of gang and the method of working. Alternatively, an efficient method of working may require a different cranage configuration to that first thought of.

In turn formwork usage will have to be considered and the concreting programme planned to allow for optimum re-use, with sufficient time allowed for any necessary alterations. This is where the contractor may look for savings, for example, by keeping column sections constant all the way up the building, because it may be that the cost of the extra material thereby required in the upper storeys will be more than saved on labour, plant and formwork.

Decisions of this kind can often only be made in the light of the actual construction programme which is envisaged, and whether site-mixed or ready-mixed concrete is to be used.

The now defunct Property Services Agency of the Department of the Environment in the UK produced an interactive computer system (COCO) which was intended to assist a cost planner in making the right cost decisions on some of these matters, although this never emerged from the experimental stage.

A further matter to be considered is the staffing of the site, and the number and type of supervisory and control staff. The contractor will also be very concerned, even at an early estimating stage, with the integration of the engineering services work with the construction programme. Some of this integration concerns the way in which the contractor and the various sub-contractors will have to work together, because the installation of pipes, conduits and components may be incorporated into the design in a way which means that they will interfere with other contractors' work sequences.

The planner/estimator will also be on the look-out for major items of equipment whose placing may cause difficulties. Equally important will be the question of whether, and if so when, the permanent engineering facilities – lighting, heating, toilets, and passenger and goods lifts in particular – may be used to facilitate the construction work.

The duration of the complete project, and of major stages in it, will have a most substantial effect on plant, supervisory and establishment costs, and this duration may in the end be determined as much by service sub-contractor requirements as by the builder's own work. In highly serviced buildings in fact they may be the prime determinant.

The criticism is sometimes made that, in assessing and balancing all the above matters, builders' estimators and planners tend to attach too much importance to optimising their 'own' part of the work, whereas in the end it is often the work of the specialists and their coordination which proves to have dictated the time for a project. This is perhaps where the independent professional cost planner can take a more balanced view.

Resource-based estimates are certainly no cheaper to prepare than a quantity surveyor's elemental estimate. Much the same sequence will therefore be employed:

- Strategic estimates taking account of major factors only, often those which are of importance in a comparison between alternative schemes. A resource-use plan at this level may also be used in the preparation of a competitive tender based upon BQs.
- A detailed production plan and estimate, probably incorporating a full critical-path network. This would be intended for use as a production control document, and would be unlikely to be prepared until the design itself had been finalised and until the contractor was expecting to be chosen to construct it. It would almost certainly involve consultation with major sub-contractors, and the use of an appropriate computer package.

Obtaining resource cost data for building work

Keeping an accurate record of costs split into:

- site labour,
- materials and sub-contractors,
- plant and
- establishment charges

and allocating them to the right contracts is comparatively easy, but the lump sums obtained do not really contribute very much to the understanding of how costs are incurred.

A detailed site costing system is much more difficult to arrange. In theory it might be possible to cost each contract in terms of finished work, attaching costs to the measured items (or groups of items) in the BQs, but this is rarely done because it is difficult and expensive to keep records of time and material for a multiplicity of items and subsequently process them.

In addition, many costs are not directly related to specific quantities of work. We can identify four types of cost, as follows:

- *Quantity related.* These are the costs which bear a straight relationship to quantity of finished work – many material costs fall into this category as do some components of labour cost. With this type of cost, twice the quantity of work costs twice as much.
- *Occurrence related.* These costs are related to a particular event, or occurrence, such as the bringing of excavation plant to the site, or the moving of a plasterer's equipment from room to room. While the scale of the occurrence is obviously affected by the scale of the work which is anticipated, its costs will not vary in proportion to actual quantity of work executed.
- *Time related.* Some costs (such as the hire of a major item of plant) are related to a length of time, and not to the amount of work done in that time.
- *Value related.* Fire insurance, for instance, will be related to the value of the project. Establishment charges are also often allocated on a basis of project value and can be included under this heading.

In addition to these four types of cost, it must be remembered that some of the work done will relate to temporary items (such as erection of site huts and conveniences) which do not form part of the permanent works. On many projects the shuttering of concrete work is a very large item in this category.

Identification of differing variable costs

It is all very well to make these rather academic distinctions between different types of cost, but in practice they may be difficult to separate. For instance the plasterers are unlikely to keep separate the time involved in moving from room to room, and the charge for a major item of plant may well incorporate several cost types, for example:

- *Quantity related*. Fuel, and part of hire charge.
- *Occurrence related*. Bringing to site and removal, moving, assembling and dismantling.
- *Time related*. Part of hire charge.

However, even an arbitrary division into these different types will be better than trying to allocate total cost pro rata to quantity of finished work, and would form a more suitable basis for using cost information for estimating or cost planning purposes.

Recent revisions of the Standard Method of Measurement have taken a welcome step towards identifying costs under these various categories.

Costing by 'operations'

In practice, costs are rarely kept to any lower level than an 'operation', which has been defined by the Building Research Establishment as 'a piece of work that can be completed by one man, or a gang of men, without interruption by others'. A typical operation might be the whole of the brickwork to one floor level, or the whole of the first fixings of joinery.

It is comparatively simple to record costs on site for such overall parcels of work, and to allocate materials to them, and it is much more meaningful to attach the time-related and occurrence-related costs to total operations than to try to split them up among quantities of measured work to which they bear little relation.

If the contractor's estimate can be similarly sub-divided it will be possible to compare cost with estimate at each stage of the project. But comparing actual costs with estimate is one of the main purposes of costing by operations; as already explained, this approach does not lend itself to the provision of cost planning data for future projects.

Use of resource-based cost information for design cost planning

The contractor would appear to be in a much better position than the consulting quantity surveyor/cost planner as far as real cost information is concerned, yet would be ill advised to use the actual costs from a particular project for cost planning another quite different one. There are two reasons for this:

- Many of the factors which affect site costs on a particular project have nothing to do with the design of the building and will not repeat from job to job. These include site conditions, the weather, industrial and personal relations, the skill with which the work was organised, accidents, late delivery of materials, defective work, failure by sub-contractors.
- Although the 'operation' is an excellent concept for collecting costs it is very difficult to re-use the information arising from it because each operation is unique. The only way in which, say, first fixing of joinery on one project can be compared with another quite different one is by looking at the quantities of the different types of work involved, and the whole essence of operational costing is that the costs are not broken down in this way. If they were, we would be back to costing work items from the BQ, and, as explained, this is impracticable. In some instances, the operation may embrace more than one cost planning element.

However, the contractor's operationally based estimates have neither of these disadvantages when considered as a data base for cost planning:

- They are based upon an assessment of 'average' costs, and as such will have a closer relationship to future tender prices than ascertained costs from a particular project.
- They will break down each operation into measurable characteristics for pricing purposes (the estimated costs of which may be based either upon experience or work study methods) and these can be used for the purposes of design cost analysis.

In so far as the contractor has access to this kind of estimate, and the consulting quantity surveyor has not, the contractor may be said to have an advantage in cost knowledge. Against this the average quantity surveyor is much more experienced than the average builder in translating project cost data into terms which are relevant at early design stage, and this expertise is essential to cost planning and control – the quantity surveyor's need is for a BQ which reflects more closely the way in which operational estimates are built up.

For instance, if the 'preliminaries' section of the BQ were always priced in a consistently itemised way (like the other sections of the BQ) the consulting quantity surveyor could analyse and manipulate this important part of the cost in detail instead of as a 'percentage of the measured work'. It must also be remembered that the proportion of project cost represented by the execution of the measured builder's work is steadily declining, and the proportion represented by specialist work, fixed charges and management costs is increasing.

Summing-up

Since all contractors pay much the same prices for labour and materials the differences in their costs are mainly to do with how well they organise and manage their projects. In order to prepare a resource-based estimate they therefore need to plan the construction, and this is almost impossible to do until some sort of drawings exist. A resource-based estimate is therefore not appropriate for the earliest stages of budgeting or planning the accommodation to be provided for the client. The contractor will need to plan the construction using a bar chart and/or a network diagram.

The proportion of total costs represented by the main contractor's direct costs is steadily declining. It is much more difficult to use the costs of a previous job to forecast the cost of a new one than it is to do the same thing with prices.

Further reading

Pilcher, R. (1994) *Project Cost Control in Construction*, 2nd edn. Blackwell Science, Oxford.
Raftery, J. (1991) *Models for Construction Cost and Price Forecasting*. RICS, London.

Chapter 19
Cost Control at Production Drawing Stage

Cost checks on working drawings

After the cost plan has been prepared (by whatever means), and agreed, it will be the cost planner's task to see that the work shown on the working drawings and measured in the BQ corresponds with the plan, as otherwise the tender figure is likely to be quite different from the amount of the estimate.

Any necessary adjustments must be made before the contract documentation, including the BQ, is finalised, not afterwards. For this purpose all drawings should be cost checked as they come into the cost planner's office; this must be a matter of routine and a rubber stamp should be made so that the drawing can be marked when the check has been done.

Unfortunately this stage in the quantity surveyor's work (the receipt of working drawings and subsequent revisions and details) is usually a hectic one and there is little time to spare for things like cost checks. Again it is unfortunate that these checks can best be done by a senior person, who is usually very busy in other directions.

Where the cost plan has been based upon approximate quantities tied to a fairly detailed specification, and where the design has proceeded in accordance with it, the work involved in cost checking should be minimal. If, however, the design has been radically changed for some reason it may be better to start the process again and prepare a fresh cost plan at as early a stage as possible, rather than attempting to check the detailed drawings against a cost plan prepared for a somewhat different building.

Cost checking procedures are quite the most time-consuming and error-prone part of the whole process and the more they can be legitimately minimised the better. Should it prove to be necessary to undertake a thorough cost check of the whole design (perhaps because the cost plan did not give the architect sufficient guidance), then the drawings will have to be prepared by elements if the feedback from the cost checks is to be of any real use.

The extent to which cost checking should be carried out will depend upon:

- *The amount of extra time which the quantity surveyor can be allowed*. This will vary from job to job, but it must be realised that if lack of time prevents essential cost checks from being carried out then the whole attempt at cost planning will largely be a waste of time.

- *The amount of apparent alteration to the scheme since the cost plan was prepared.* If the cost plan was based upon a very similar project for the same architects and client and the details of specification and design were therefore known fairly well, the cost check may entail little more than a quick glance at the drawings to see that nothing is substantially different. Even where the circumstances are not quite as ideal as this it may still be possible for the quantity surveyor to be satisfied by a quick inspection that the drawings show what the cost plan envisaged. The drawings should be stamped and initialled, even so.

- *The amount of detail in the cost plan.* If it was possible to take out fairly full approximate quantities the cost check may simply consist of comparing these quantities with the working drawings and checking that the specification is unaltered.

- *The degree of confidence which exists between cost planner and architect.* If the architect is known to be cost conscious, capable of and enthusiastic about cost design, the cost checks will be much less important than where this confidence is lacking.

- *The familiarity of the type of project.* Most cost planners would feel fairly confident of cost planning a school, whereas a planetarium would be a very different proposition. In the latter case the cost plan would probably incorporate rather a lot of assumptions and it would be necessary to check these in detail.

- *The importance of the element.* It would be a very self-confident cost planner who would not bother to check the external walling, or the roof of a single-storey building, even under the most ideal conditions. On the other hand, if time is pressing some of the smaller elements can often be ignored.

Carrying out the cost check

Again this is a suitable occasion for using standard forms. Where the architect is designing by elements the procedure will be considerably simplified, but even if this is not the case the check must still be done by elements.

The design of the cost check form may be left to the tastes of the firm or other organisation. It should show at least the following information:

- Number of check. The checks on a particular project should be numbered consecutively, starting at 01.
- Total estimated cost of project after completion of previous check. (This amount will be got from the previous cost check form, or if this is the first check it will be the unaltered total of the cost plan.)
- Reference number of drawing(s) or other information being checked.
- Element being checked.
- Amount allowed in cost plan in respect of element (or as amended by previous checks).

- Estimated cost of element as calculated from the drawing(s), or other information, being checked.
- Difference (plus or minus) between the two last.
- Estimated cost of project after this difference has been added to or deducted from the previous estimated project cost.

A typical example of a cost check form is shown in Fig. 19.1. As each cost check is carried out a copy of the cost check form should be sent to the architects so that they have an up-to-date running total of the project. If any check shows a substantial increase or decrease it would be as well to discuss the matter with the architects rather than merely sending the results of the check to them. Similarly, they should be warned if the general standard of detailing appears to be more lavish than was allowed for.

CONTRACT: Office Block, Midtown
COST CHECK No 3
ELEMENT: Structural Frame

DATE: 06.07.99
Gross internal floor area: $2390\,m^2$
Total cost of project forward from cost check No 2: £1,907.220

			£ total	£/m^2
Total cost of element from updated cost plan			125,600	52.55
Elemental cost check:				
(see attached dimensions)				
$380\,m^3$ Reinf conc in beams and cols	£90	£34,200		
$2950\,m^2$ Formwork to beams and cols	£16	£47,200		
31 tons Reinfct (as Mr Smith of Consulting				
Engineers 28.06.99)	£1200	£37,200		
Sundries		£10,000		
Revised cost of element carried forward		£128,600	128,600	53.80
Amount of saving/extra:		+ £3,020	£1.26	
Revised total cost of project carried forward to next cost check:	£1,907,220 + £3,020 = £1,910,240			

Fig. 19.1 Example of cost check form.

If a drawing shows changes in more than one element a separate form should be used for each. The rough workings in connection with each check can be done on dimension paper and stapled to the office copy of the form.

A list should be kept of all drawings or other information received, with columns for marking:

- That the cost check has been carried out.
- The reference numbers of the cost check form or forms.

- The elements involved. The purpose of this is to enable the checker, on receiving a drawing showing (say) part of the roof to look back through the list to see whether any adjustments have already been made to this element.

If it appears that a detailed cost check of any particular drawing or information is not necessary the list can be marked accordingly. However, all drawings and information (such as sub-contractors' quotations or replies to queries) must come to the cost checker in the first instance, even if someone else is waiting for them.

In connection with cost checking it is as well to remember the total cost of the project and also the inherent margin of error in cost planning. A cost planner whose estimates consistently get within plus or minus 5% of the accepted tender is doing well. In these circumstances it is obviously not worthwhile to spend much time cost checking isolated details of a large project, as the possible differences in cost are so small compared to the probable overall margin of error.

Prices for cost checking may be obtained from the same sources as have been used earlier. But as there will be a tendency for the items to be more detailed than at cost plan or estimate stage the cost planner is likely to be using either built-up rates, price book rates or actual rates from the BQ from which the example analysis was prepared. Again it must be remembered that the contractor's permission must have been obtained for this latter course.

It will be very tempting to lift prices from BQs for other projects at this stage, but the errors that can arise from this practice have previously been pointed out. If prices are built up or obtained from price books the level of market prices on which the estimate is based must be remembered. It will also be necessary to include a realistic percentage for preliminaries and insurances; this will not necessarily be the same percentage as the contractor showed in the analysed example.

Use of an integrated computer package at production drawing stage

Figures 19.2a to 19.2d illustrate an extract from the output of an integrated computer cost planning package (CATOpro) relating to an urban site development for a housing association.

Perhaps the biggest advantages of a computer package such as this is that the figures for the total project are automatically updated whenever a change is made to any element or part element, without the opportunities for error or omission which are always present with a manual system.

Figure 19.2a shows a revision to the Substructure element of sites A, B and C. This is very similar in appearance to the manually produced example for the structural frame of the Midtown office block in Fig. 17.4 (Chapter 17). The input to Fig. 19.2a is the measurement of approximate quantities from the drawings, either manually or entered by digitiser, or measured and calculated using a lower level of the computer package (not shown here).

Figures 19.2b and 19.2c (level 2) show the next stage, the revised elemental summary page for sites A, B and C. The various figures can either be inserted directly on an elemental basis per element or automatically derived from a level 3 cost check. In this case the figures for '2.0 Substructure' have been transferred by the computer from the level 3 elemental cost check which was shown in Fig. 19.2a.

Figure 19.2d (level 1) shows the updated summary of cost for the whole project, automatically updated from the more detailed levels of working of which examples have been given. The figure for sites A, B and C has been transferred by the computer from the level 2 elemental summary (Fig. 19.2c).

It will be noted that costs are given in £/ft^2 as well as in £/m^2, this being the unit of measure preferred by most real estate clients.

Use of resource-based techniques at production drawing stage

This technique has been introduced in the previous chapter. If this approach is being used a detailed production plan and estimate would be prepared at this stage, probably incorporating a full critical-path network. This would be intended for use as a production control document, and would be unlikely to be prepared until the design itself had been finalised and until the contractor was expecting to be chosen to construct it. It would almost certainly involve consultation with major sub-contractors, and the use of an appropriate computer package.

Cost reconciliation

If tenders have been called for and received the tender which is most likely to be accepted should be reconciled with the cost plan. If, as should occur, there is little difference between the totals this is still worth doing; there are quite likely to be large compensating discrepancies in the various elements and these may provide information for future use.

The reconciliation is in fact a comparison of the final cost plan (as amended by cost checks) with a cost analysis of the tender. This also ensures that the cost analysis will be done at the earliest possible moment instead of being left until 'someone has time to do it'.

There may be all sorts of explanations for a considerable difference between the cost plan estimate and the actual cost of a particular element; perhaps the cost planner made a mistake, quite possibly the builder's estimator made a mistake or did a deliberate price adjustment. However, if there is a definite one-way divergence between cost plan and tender right through all the elements it is likely that either the cost planner misjudged market levels or that the tenders themselves are abnormally high or low. If the cost planner feels that the latter is the case then this could be pointed out.

Note that permission is never required to analyse the contractor's tender for cost reconciliation purposes. It is only if the analysis is to be published, or if the individual prices are to be used for cost checking other projects, that the question arises.

URBAN SITE REDEVELOPMENT
SITES A, B AND C
2.0 SUBSTRUCTURE

HOUSING
ASSOCIATION

	Description	+	Quantity	Unit	Rate £	Calc	Total £	Notes
1	Clear/strip site		10587.00	m²	2.00		21,174	Site layout
2	Strip foundations		2061.00	m²	60.00		123,660	To houses
3	Extra over allowance for stepped foundations		1.00	Item	3000.00		3,000	
4	Piles, ground beams and pile caps			m²	105.00		17,325	To flats
5	In-situ reinforced conc ground slab with dpm and screed		2226.00	m²	60.00		133,560	Assumed 300 thick
6	Allowance for cut and fill earthworks		1.00	Item	25000.00		25,000	
7								
							323,719	
U/D	4,350/M2							

Fig. 19.2a   Bloggsville United Training Centre – revision to Substructure element of sites A, B and C.

Elem code	Element	%	Cost £/m²	Cost £/ft²	Quantity	Unit	Rate £	Sub total	Total £	Notes
1	1.1 DEMOLITIONS									In Enabling Works
2										
3	**TOTAL**		—	—	—		—	—	—	
4	1.2 ALTERATIONS									In Enabling Works
5										
6										
7	**TOTAL**		—	—	—		—	—	—	
8										
9	2.0 SUBSTRUCTURE	10.84	74.42	6.92				323,719.00	323,719	
10										
11										
12	**TOTAL**	**10.84**	**74.42**	**6.92**	—		—	**323,719.00**	**323,719**	
13	3.0 SUPERSTRUCTURES									
14										
15										
16	3.1 ROOF	6.77	46.46	4.32				202,098.00	202,098	
17	3.2 EXTERNAL WALLS	10.51	72.14	6.70				313,805.00	313,805	
18	3.3 UPPER FLOORS	4.51	30.97	2.88				134,728.00	134,728	
19	3.4 WINDOWS AND EXTERNAL DOORS	12.18	83.61	7.77				363,725.00	363,725	
20	3.5 INTERNAL WALLS AND PARTITIONS	6.42	44.08	4.10				191,741.00	191,741	
21	3.6 INTERNAL DOORS	3.57	24.51	2.28				106,600.00	106,600	
22	3.7 STAIRS	3.06	21.03	1.95				91,500.00	91,500	
23										
24	**TOTAL**	**47.02**	**322.80**	**30.00**	—		—	**1,404,197.00**	**1,404,197**	
25										

URBAN SITE REDEVELOPMENT SITES A, B AND C — **HOUSING ASSOCIATION**

Fig. 19.2b Revised elemental summary page for sites A, B and C (level 2).

URBAN SITE REDEVELOPMENT SITES A, B AND C

HOUSING ASSOCIATION

Elem code	Element	%	Cost £/m²	Cost £/Ft²	Quantity	Unit	Rate £	Sub total	Total £	Notes
26	4.0 INTERNAL FINISHES									
27										
28	4.1 WALL FINISHES	6.34	43.49	4.04				189,177.00	189,177	
29	4.2 FLOOR FINISHES	4.33	29.72	2.76				129,268.00	129,268	
30	4.3 CEILING FINISHES	2.48	17.03	1.58				74,097.00	74,097	
31										
32	**TOTAL**	**13.15**	**90.24**	**8.38**	—		—	**392,542.00**	**392,542**	
33	5.0 FITTINGS AND FURNISHINGS									
34										
35										
36	5.1 FITTINGS	5.42	37.17	3.45				161,700.00	161,700	
37										
38	**TOTAL**	**5.43**	**37.17**	**3.45**	—		—	**161,700.00**	**161,700**	
39	6.0 SERVICES INSTALLATIONS									
40										
41										
42	6.1 MECHANICAL SERVICES	13.11	90.00	8.36				391,500.00	391,500	
43	6.2 ELECTRICAL SERVICES	9.78	67.11	6.24				291,910.00	291,910	
44	6.3 BWIC SERVICES	0.67	4.60	0.43				20,000.00	20,000	
45										
46	**TOTAL**	**23.56**	**161.71**	**15.03**	—		—	**703,410.00**	**703,410**	
47										
48										
U/D	**4,350/M2**	**100.0**	**686.34**	**63.78**				**2,985,568.00**	**2,985,568**	
									2,985,568	

Fig. 19.2c Revised elemental summary page for sites A, B and C (level 2).

URBAN SITE REDEVELOPMENT							HOUSING ASSOCIATION	
Summary	Cost	Calc	Total £	Area	Cost £/m²	Cost £/Ft²
1 ENABLING WORKS	77,231		77,000			
2 SITES A, B AND C	2,985,568		2,986,000	4,350	686	64
3 EXTERNAL WORKS ASSOCIATED WITH UNITS	288,216		288,000	4,350	66	6
4 PRELIMINARIES	327,300		327,000	4,350	75	7
5 INFRASTRUCTURE	809,084		809,000			
6 CONTINGENCIES	226,660		227,000			
Project	4,714,059		4,714,000			

Fig. 19.2d Updated summary of project cost (level 1).

Completion of working drawings and contract documentation

The completion of the working drawings and the contract documentation will normally mark the end of the cost planning process as such, although the vital task of cost control will have to continue throughout the project to ensure that the planned cost is achieved.

A critical assessment of elemental cost planning procedures

It has been suggested that elemental cost planning techniques will be justified if they consistently succeed in forecasting cost within a margin of 5% up or down, and indeed more recent research suggests that even this target may be over-optimistic, given the variability which exists in tender levels. It might well be asked whether all this work is worthwhile if, at the end of it all, the tender is liable to differ from the estimate by such an amount. Simple single price rate methods of estimating have often come far closer than this and involve much less work.

The answer is threefold:

- Most budgets are flexible by at least this percentage or, if they are not, it is simple enough to make the requisite modest alterations in the scheme.
- It must be emphasised that traditional single rate methods cannot be relied upon to achieve anything like this standard of accuracy, and that for every 'spot-on' square metre estimate there is another instance of such an estimate being up to 50% out. But where an estimate, however prepared, forecasts the cost to within 1% or 2% under traditional competitive tendering conditions, then luck as well as skill will have played its part in the result.
- Elemental cost planning achieves a balance and economy in the building which cannot be attained by any other method using traditional tendering procedures. A real-life example concerns a technical school which was estimated on a square metre single price basis and where the difference between estimate and tender was much less than 5%. Satisfactory though this was, nobody was able to answer a very simple question from the architect during the design stage as to whether that estimate would cover a certain type of wall cladding. There was nothing in the estimate with which it could be compared.

In view of the alleged unreliability of estimates prepared by single rate square metre methods it is as well to explain again why these methods can safely be used for the preliminary estimate before the cost plan is prepared. It is because the cost plan will show up any error in the estimate before working drawings or BQs have been prepared, and the necessary adjustments can be made before any of the detail work is started. This is a very different situation to discovering the error at tender stage when time and money have been spent designing and tendering for an impossible scheme in detail.

While elemental cost planning has made an outstanding contribution to the study, forecasting and control of building costs, it nevertheless suffers from a number of basic limitations which have prevented it from developing much beyond the stage which it reached within the first few years. One might wonder whether it is doomed to be like the lead–acid electric battery which, although very useful for over a century, has never been capable of development into the kind of power-storage unit that the world is waiting for.

The limitations in question are:

- Nobody has yet produced a set of elements each of which performs a single function but which can be easily cost-related, nor are they ever likely to.

This means, amongst other things, that it is not really possible to compare the cost performance of the same element on two different buildings, nor, within one element, to compare two different technical solutions concerning one building. In order to attempt this task (which the quantity surveyor is often asked to do) it is necessary to consider all sorts of extraneous matters, some of which are difficult to quantify. It might be thought that it would be possible to overcome this limitation by the use of sophisticated computer programs which would enable the interaction of the elements to be worked out exhaustively. This might well be possible, but it would be expensive and at the present time is not worth doing because of the next limitation.

- As we have already seen, but as is worth repeating here, the 'costs' which are being manipulated may bear no relation to fact; they are not really 'data' in any scientific sense.

It is a waste of time and money to use sophisticated methods to manipulate inaccurate data, as is recognised in the computer world where the maxim 'garbage in, garbage out' is well known. In fact, such processing is worse than useless, because the fact that sophisticated analysis methods have been employed leads people to attach undue importance to the results. They would immediately recognise the original data as unreliable if it were presented to them in an unprocessed form.

It must be remembered that there is no pressure on contractors to insert rates in the BQ which represent carefully estimated production costs for each of the items, and there are many reasons (as set out previously) why they do not. The rates in the BQ are not 'retail' prices for which the contractor is offering to execute individual pieces of work in isolation, but are merely a notional break-down of the total offer.

However, even if it were possible to compel the contractor to show true cost estimates, and to enforce standard practice in defining and allocating overhead costs, there would still be the difficulty of setting production costs against square metres of design elements, as so many of these costs (e.g. the tower crane) are not directly related to element unit quantity.

- Even if the two previous limitations could be overcome the whole system suffers because under normal conditions there is no contractual commitment to the cost plan.

The QS has no responsibility for the tender amounts and the contractor has no interest in the cost plan. The exercise thus has a great deal in common with weather forecasting. The cost planner is able to draw on a considerable body of past experience and can consider present trends, but is in no position to guarantee the result. The forecast will often be right, but if any unusual circumstances manifest themselves it can still be badly wrong.

The comparatively crude methods which have been set out in this and preceding chapters are usually adequate for forecasting within the wide limits of market pricing; any attempt to get much closer in these conditions is unlikely to be a worthwhile use of resources. Therefore if the cost planner cannot offer any kind of guarantee to the client the latter is unlikely to agree to expenditure on expensive and sophisticated methods of 'control' which are not in fact controlling anything but are just hopeful forecasts.

If a higher standard of cost control is required than is provided by the reasonable development of these traditional methods then it will be necessary to sacrifice some measure of market freedom to obtain it, for example by involving the contractor at an early stage. There is just no way round this.

Summing-up

- Elemental cost planning should ensure that the tender amount is close to the first estimate, or alternatively that any likely difference between the two is anticipated and is acceptable.
- Elemental cost planning should ensure that the money available for the project is allocated consciously and economically to the various components and finishes.
- Elemental cost planning does not mean minimum standards and a 'cheap job'; it aims to achieve good value at the desired level of expenditure.
- Elemental cost planning always involves the measurement and pricing of approximate quantities at some stage of the cost plan or cost check.

Chapter 20
Methods of Procurement

Meeting the client's needs

'Procurement' is the term used to describe the total process of meeting the client's need for a building, starting at the point where this need is first expressed. We have to remember that the easiest method of procurement is for the client to buy or lease suitable accommodation which already exists, in which case the services of an architect, cost planner or building contractor will not be needed at all! However, if the client decides on a new building, or needs to refurbish an old one, it will be necessary to enter into contractual relationships with one or more people or organisations inside the construction industry in order to get it designed and built. This applies even if the client happens to be a firm which employs its own building labour for maintenance, etc. – it is most unlikely to have the full range of needed skills and resources in-house.

Standard Forms of Building Contract

Until the passing of the Housing Grants, Construction and Regeneration Act 1996, building contracts and sub-contracts were governed only by the general law of contract. Early in the twentieth century the Royal Institute of British Architects (RIBA) had published a model form of building contract which gained wide acceptance. Building contractors, however, felt that a contract drafted unilaterally by an organisation, however well meaning, which represented only one side of the industry was inherently unfair.

An organisation called the Joint Contracts Tribunal (JCT) was set up to include contractors' representatives. The JCT took over administration and revision of the RIBA Form, and was gradually expanded to include representatives of all the bodies involved in procuring a building including sub-contractors and clients. It now publishes a whole range of forms to suit almost every conceivable kind of contract. The JCT forms do not have any kind of statutory authority, and there has never been anything to prevent clients or contractors from either using their own forms or altering the wording of the JCT forms to suit themselves.

Although in theory a contract is mutually agreed between its parties, in practice a building contract or sub-contract is usually presented by one party to the other on a pre-printed 'take-it-or-leave-it' basis. This led to the imposition of unfair terms where either the client or the contractor was in a dominant commercial position, or had much greater experience than the other party.

The Construction Act 1996

Because of concern at the increasing abuse of this situation the Construction Act 1996 (of which the full title is the Housing Grants, Construction and Regeneration Act 1996) laid down those matters which a building contract must cover, including:

- methods and timings of payment;
- fair means of settling disputes.

In the absence of satisfactory clauses in the contract dealing with such matters the provisions of the Scheme for Construction Contracts laid down in the Act will apply. If there are any unfair provisions in the contract which conflict with the Scheme, the Scheme will automatically over-ride them.

Basically the Act applies to all normal building contracts except a contract made directly with an existing or intending residential occupier. Its main effect as far as cost control is concerned is to make it very difficult to place unusually onerous obligations on builders or sub-contractors, particularly infamous 'pay-when-paid' clauses.

Basic forms of building contract

Although there are many different forms of contract (including all the JCT variations) they are basically of two different kinds:

- *Cost reimbursement.* The contracting firm agrees that all its expenditure on labour, materials, etc., will be met by the client, on top of which it will charge a fee on an agreed basis (e.g., 'I will redecorate your drawing room for the cost of paint plus £10 per hour for my time').
- *Price in advance.* The contracting firm agrees to carry out its obligations for a sum of money agreed in advance (e.g., 'I will redecorate your drawing room for two hundred pounds').

The two types of contract are illustrated with 'homely' examples because they are rarely used in their pure form on large projects, for two main reasons:

- The lump sum 'price in advance' contract involves too great a risk to the contractor (the building may need deeper foundations than was bargained for, or inflation may send wages rocketing).
- The cost reimbursement contract encourages waste and extravagant working by the contracting firm, which has not got to pay for anything that it uses. In fact, in the most primitive form of cost reimbursement contract where the

profit is a percentage of the cost, the contractor has a positive incentive to waste as much money as possible.

From the point of view of cost planning, however, there is rather an interesting paradox:

- The 'price in advance' contract gives the client almost no control over the details of methods, programming or expenditure, but gives an excellent forecast of the total cost.
- The 'cost reimbursement' contract, on the other hand, allows the client to give orders about the tempo or methods of work, and to obtain detailed allocations of actual site costs, but the total cost can never be forecast with the same degree of certainty.

We can summarise the situation as follows:

	Price in advance	Cost reimbursement
Requirements must be fully known and defined in advance of work	Yes	No
Binding cost commitment	Yes	No
Management involvement by client team possible	No	Yes
Production cost feedback possible	No	Yes

It should also be noted at this stage that the type of building contract which is chosen will have some effect on the type and scale of professional advice which is employed. For example, in the case of a large 'cost reimbursement' contract it would almost certainly be necessary to employ a quantity surveyor to check all the invoices and wage payments.

Building contracts in detail

We can now look at the various contractual methods of building, trying as far as possible to match them against the client's time and cost criteria which were set out in Chapter 4. While this may suggest the best approach to the individual problem it must be remembered that no form of contract is proof against things going wrong, and a keen combination of design team and building team will make a good job of things whatever the contractual arrangements. However, all other things being equal, a suitable form of contract will help.

Cost reimbursement

Cost plus percentage

The advantages of cost plus percentage are:

- It is the most convenient contractual basis of all.
- The contractor can be selected, the contract placed and work started before

the scheme has been finalised and without any estimates or quantities needing
to be prepared.

- The contractor's management methods, in theory at any rate, can be used for
  the direct benefit of the client.
- The client also knows that the contractor will not make an exorbitant profit.

However there are disadvantages:

- The usual cost reimbursement drawbacks of poor cost forecasting facilities
  and low productivity arising from the fact that people are not 'working to a
  price' so can afford to do things properly (or slowly and extravagantly).
- There is a positive incentive towards wastefulness because the contractor is
  being rewarded with a percentage of everything that is spent.

Because of these drawbacks it is not surprising that this type of contract has a
bad cost reputation. However, because of the virtues which have been mentioned
it is a widely used method for such projects as:

- emergency first aid and repair work;
- alterations and repairs to old buildings where the extent of the works cannot
  be foreseen until the contract has started;
- contracts where very high quality work is required;
- contracts where cost may be important but where the client above all wants
  control over the method of working;
- contracts where a good long-term relationship exists between the client and
  the contractor.

Cost plus fixed fee

This is an attempt to mitigate the worst feature of the above type of contract by
paying the contractor a fee based on the estimated cost of the project instead of a
percentage of the actual cost. This has the advantage (compared to cost plus
percentage) of reducing the incentive for wastefulness. However, there are two
disadvantages:

- A fairly detailed scheme and an estimate have to be prepared before work can
  start, so losing some of the advantage of the 'cost plus' system.
- It is likely in practice that the so-called fixed fee will have to be renegotiated
  at the end of the job because of the major variations which are sure to arise in
  projects of the type for which such a contract would be used. This is likely to
  be based upon the contractor's actual cost, so effectively returning to the cost
  plus percentage system.

Target cost

This involves an attempt to get the benefits of cost reimbursement without the
disadvantages. A BQ is prepared and priced by negotiation to arrive at a target
cost; the contractor then carries out the work on an actual cost basis.

There are various methods of arriving at the final price:

- If the actual cost is lower than the target cost the contractor and client usually split the difference between them on a pre-arranged basis.
- If the actual cost is higher than target the contractor has to be content with the actual costs plus a small overhead percentage.

The advantages are:

- The contractor has an incentive to do the job as cheaply as possible and the client gets a direct benefit if this is achieved.
- The system retains many of the benefits of cost reimbursement, especially the ability of the contractor and client to work closely together in the management of the project.
- It can be used with particular success on a 'continuation' basis where the client has a succession of projects in which the same contractor can participate.

The main disadvantages of target cost are:

- The target cost is subject to revision in respect of the many likely variations (so that one might as well have an ordinary lump sum or remeasurement contract).
- The quantity surveyor's fees are likely to be high because there is dual documentation (cost reimbursement accounts to be checked and BQs to be prepared, priced, agreed and updated).

Management-based projects

A method of procurement whereby a management team organises the project on the client's behalf, for a fee. It was first used with success on major industrial projects where time and integration with complex engineering contracts were vital. It has since been increasingly used for more general projects and particularly in refurbishment works, where close relationships between the client's representatives and the construction team are very important.

There are three different approaches which may be used:

- management contracting,
- construction management and
- project management

and these, together with their advantages and disadvantages, are dealt with at greater length later in the chapter.

Direct labour

This is not, strictly speaking a 'contract' at all, because the client employs labour, buys materials and engages sub-contractors in a series of minor contracts, doing all the organising and bearing all the risk. It is not often used except by clients who normally maintain a building department, such as major industrial concerns and public authorities.

This method is subject to the usual cost reimbursement advantages and disadvantages, although it is sometimes possible to set up a 'model system' of tendering and cost control. A particular advantage is that, because the client organisation is dealing directly with its own people and suppliers, communications can be simple and informal. In addition, the people doing the work should be acting in the client's interests.

The disadvantages of this method are that a client for a one-off building would have great difficulty in getting together an efficient management team on such a short-term basis. In addition, in the public sector direct labour has suffered from being the subject of strong party political controversy, and there have been a number of notorious cases of corruption and inefficiency.

Price in advance

This is probably still the most widely used type of contract, in spite of various difficulties. One of these difficulties is, as we have seen, inflation. Because large building contracts will last for a year or more it is unreasonable to ask a contractor to bear the total risk of inflation during this period. Therefore most of the contract methods set out hereafter have two versions:

- One for use on small to medium projects in times of economic stability where the contractor bears the inflation risk.
- One for larger projects or at times of economic instability, where the risk is borne in whole or in part by the client.

It should be noted that while the version where the contractor bears the risk gives the client a firm cost forecast the alternative approach may well be the cheaper, even on short-term jobs. This is because contractors do not like this risk, and tend to overprice it if they are given a choice of tendering on the two methods. We have seen from the cash flows in Chapter 17 why the contractor is in a worse position than the client to take even a modest degree of extra risk.

There are various types of price in advance contracts, which will now be considered.

Lump sum

- In a lump sum contract the building firm commits itself to a price on the basis of the work shown on the drawings and in the specification.
- Although commonly used abroad, it is difficult to obtain prices in competition on this type of contract in the UK, except for very small schemes.
- A quantity surveyor is not usually employed and BQs are not used; this means that preliminary cost forecasting, budgeting and valuing of changes in the scheme may be unsatisfactory.

- However where the client's requirements are firm, and the drawings and specification are good, this is a very simple and effective contractual basis for modest sized projects.

Contract based on BQs

In spite of discouragement and disparagement in the three major reports on the construction industry which were produced between 1988 and 1998 – Building Britain 2001, Latham and Egan – this is still the normal mode for carrying out major building work in the UK by one-off clients. It will therefore have to be considered in most detail. Wide usage has developed a number of variations on the basic theme, as follows.

Competitive tender – JCT Standard Form of Contract with BQs:
- Most major work in the UK is undertaken on this basis. The scheme is designed, and a number of contractors are asked to submit lump sum tenders based upon the pricing and totalling of a BQ prepared on behalf of the client.
- The successful contractor's tender BQ then becomes the instrument for financial administration of the project, so that the one document provides a simple means of contractor selection, price commitment, and contract management.
- This method of contract probably gives the lowest price of any, under UK conditions.
- However, although superior to any of the previous methods in regard to cost forecasting and budgeting, it has a number of disadvantages in this respect which have been set out in Chapters 12 and 13.
- It is this type of contract, with its competitive BQ rates, which provides most of the raw data for elemental cost analyses, and indeed acts as a control against which the cost of buildings, erected under less competitive arrangements can be judged.
- In practice, the provisions of the JCT Standard Forms of Contract (which attempt to be fair to everybody) make the total price and time commitments rather less firm than they appear to be in theory.

Competitive tender based on Bill of Approximate Quantities: Used as an alternative where the scheme has not been fully designed, the BQ being only a notional representation of the finished product. The work is measured as executed, and a final price agreed using the items and rates in the BQ as far as possible. This method has the advantage of permitting an earlier start to be made, however, control, both of time and money, is much weaker than in the previous case.

A contracting firm which has underestimated the cost of its commitments when tendering is sure to be able to find some way of recouping its actual costs under such a nebulous arrangement as this.

Negotiated tender: A contractor is selected early in the design stage, either with a pin or else by means of a simple tender document in which several firms are asked to state their management costs, on-costs, and labour charges. As the design develops the contract BQ is prepared and priced jointly with the contractor, who will normally be involved in the detailed design also.

Advantages include the following:

- The system assists cost forecasting and budgeting, as there is some contractual commitment to the cost assumptions made during design.
- An early start can also be made on site.
- Because everything is negotiated, the contract terms and the arrangement of the BQ can be tailor-made to suit the actual requirements of the parties.
- This method of contracting is very suitable for use on a 'continuation' basis, where the same client, professional advisers and contractor can establish a good working rapport.

Disadvantages are:

- It is unlikely to produce 'rock-bottom' prices.
- It requires some degree of special expertise on both sides to obtain the best results.

For this last reason, the quantity surveyor may well be the most influential member of the professional team and may be appointed first.

Serial contracting: This method is only open to clients with a continuing work programme.

Contractors are asked to price, in competition, a typical BQ, on which basis they agree to carry out any similar work which the client may require over a fixed period of possibly 2 or 3 years. Usually the contractor is asked to take the rough with the smooth, and to quote average rates which will apply on both difficult and straightforward projects.

An outstanding advantage is:

- Superb potential for cost forecasting, as contractually binding detailed prices are available for the future schemes before they are even designed.

But disadvantages are:

- The system relies heavily on mutual trust and it only works where the contractor has confidence in the integrity and fair play of the client, for which reason it has mainly been used in the public sector.
- As already stated, it does not usually produce 'rock-bottom' prices.

Schedules of prices

Sometimes called a measured term contract, this method is used mostly for repair and maintenance work, particularly by government departments and local authorities. It has something in common with the serial contract, as the contracting firm binds itself to a 'price list' at which it will undertake work as required for a period of years, subject to an agreed discount or premium as the case may be.

The most commonly used version is the *National Schedule of Rates for Building Works* (seventh edition 1995) which is published by HMSO. This type of contract is sometimes used as a matter of convenience for small new works.

Design-and-build

Under this system the contractor acts directly for the client, filling the roles of both the professional design team and builder.

The big advantage is that the contracting firm will be willing to commit itself firmly to a price and completion date at an early stage. This is because it knows it will have total responsibility for working out the scheme without interference from others.

Disadvantages are:

* The drawbacks of production-oriented design.
* The lack of competition. Even if two or more tenders are obtained it is difficult to evaluate the 'best buy', as what is being offered by each tenderer will be different.
* Loss of control by the client over the end product if the offered price is to be maintained.

Because of the lack of a competitive tender the client usually appoints a quantity surveyor to keep an eye on costs. The client may also ask a consultant architect to advise on design. However, the latter role is so limited and frustrating that not every architect is keen to take it on.

The design-and-build firm may have its own in-house design team or may employ outside consultants; as these latter are responsible to the design-and-build firm and not to the client this is not much more than a domestic detail.

Until 1989 design-and-build firms had an advantage compared to those using traditional procurement methods for new projects because the design costs, being part of the building contract (unlike an independent architect's fees), were not liable to VAT. With subsequent changes in VAT legislation, under which most non-housing building work is subject to VAT this advantage has largely disappeared.

Although design-and-build has been dealt with here as a price-based method there does not seem to be any valid reason why such an organisation could not be employed on a cost-reimbursement basis where the project does not lend itself to a price-based contract.

Design-and-refurbish is now becoming quite prevalent, but if carried out on a price basis it does involve considerable risk to the contractor.

Managed projects

In the early 1990s management-based projects enjoyed considerable popularity, because of the advantages they seemed to hold, such as:

- contractual relationships based on collaboration instead of confrontation;
- availability of the contractor's construction expertise to the design team.

But experience has led to some reversion to more traditional methods, partly because of a number of high-profile failures together with concerns over the forms of contract which are available.

Management contracting

Of the three approaches to managed projects mentioned above, management contracting is the closest to the traditional practices and structure of the contracting industry. In its most basic form it is little different to the old prime cost plus fixed fee contract, in spite of its high-sounding name, although its protagonists always deny this.

The main difference in theory is that the managing contractor firm does not undertake anything other than site management with its own resources, all direct works being sub-contracted by it. However this is now so much the general pattern of the industry that it is not easy to discern much real difference in practice.

Construction management

This is a development of management contracting, where the construction management firm acts in a role more like that of a professional consultant than a building contractor. All the works contractors enter into direct contracts with the client rather than being traditional sub-contractors.

This type of project is rather less attractive to building contracting firms because:

- It only uses one part of their organisation – management – which is greatly needed within their own business.
- It denies them the benefits of the project cash flow.

Conversely it does enable consultancy firms such as architects and engineers (and quantity surveyors), who could not undertake management contracting, to offer management services to clients. This practice is as yet more common in the USA than here.

The advantages claimed for both the management contracting and construction management approaches, compared to the traditional standard JCT type contract, are considerable:

- Fast-tracking, i.e. making the design and construction processes concurrent instead of consecutive, becomes possible.
- Construction-based expertise becomes available to the designer at an early stage.
- The collaboration of client and designers with the builder, instead of the usual confrontation, is especially valuable on very large and complex time-sensitive projects or in refurbishment work.

These advantages have not always been realised in practice, because:

- Traditionally trained contractor staff (and contractor organisations) find difficulty in adapting their methods to a new client-oriented role.
- The removal of adversarial relationships only occurs at the top management level. Down where the action is the direct works contractors are normally employed on the usual lowest-price-for-the-work basis and the usual confrontations apply. It is sometimes said that these are worse than usual, because the trade contractors feel that the builder, who used to be a fellow sufferer, is now insulated from cut-throat tactics whereas they are not.

However, the advantages of management contracting and construction management have undoubtedly led to their employment on large and difficult projects where any system would find itself in trouble. On such projects traditional methods would probably have got into even greater difficulties (even if it had been possible to use them at all). However, for all their new-found status, the management contractor and the construction manager are still acting largely in the builder's traditional role. In particular they are dependent for their success upon satisfactory performance by the architect, consulting engineers and quantity surveyor, none of whom they control. Their ability to guarantee the client an on-time and on-cost project is therefore limited.

Sometimes therefore the management contractor may seek to control the designers, particularly as regards the timing of information flow. But this is an unhappy compromise and the client who is looking for a manager who will accept total responsibility for performance would do better to consider a full project management service.

Project management

This is a term which sounds impressive, so it is often used very loosely. In its true sense project management involves a manager who comes in right at the top, interposed between the client and all consultants, including not only the building and engineering professionals but also those concerned with development advice and marketing.

The project manager obtains advice and services from all these, and in turn advises the client.

- The project manager is paid a fee, but does not usually handle any other money.
- All consultants and contractors have direct contracts with the client although they take their orders from the project manager.

It can be seen that the project manager is a true professional, in the sense of having no entrepreneurial interest in the project, and may come from almost any discipline. It is the ability to manage that is important – the project manager can get all the technical input that is needed from the client's consultants.

Given a good management reputation, however, there may well be a tendency for a client to pick a project manager whose background is related to those aspects which are considered to be the key ones on the development in question:

- a general practice surveyor if marketing is paramount;
- an architect where quality of design is the vital factor;
- a QS where overall cost is important.

The existence of a project manager does not necessarily imply that a non-traditional method of construction procurement will be used. It would be quite possible to use a general contractor appointed on BQs for instance. But certainly it is more common for project managers to adopt a management contracting or construction management approach, if only because the type of project that demands a project manager usually lends itself to such methods.

Managed projects generally

It has to be remembered that at the end of the day total costs are likely to be greater with managed projects than with more traditional methods. Apart from anything else, there is usually an extra level of management to be paid for. Also in the UK the traditional lump sum/BQ approach brings two major benefits which are lost when newer management methods are employed:

- All procedures are so well known that the parties do not have to go through a learning process (at the client's expense) when they are running the project.
- The provisions of the standard forms of contract are well tried in case law, unlike remodelled or 'bespoke' forms.

So it is probably a mistake to employ management methods on moderately sized straightforward jobs which are in no particular hurry.

Finally, it is of the essence of professional project management, either at construction level or at total scheme level, that the management organisation must not carry out any of the other professional or construction tasks for which it is responsible to the client.

A management contracting firm should not do any of the building work. There are some very grey areas at present involving works contractors which are group subsidiaries or group associates of the management contracting firm. Also a quantity surveying firm that is acting as project manager should not do the cost planning themselves. Similarly where project management is in use the project manager should not also attempt to act as construction manager – the roles are quite different. It is perhaps for this reason that the role of scheme project manager has not often proved attractive to construction firms, in the UK at any rate.

The usual cost reimbursement advantages and disadvantages apply to all managed projects, although the management team will almost certainly let out most of the work to sub-traders and specialists on a 'price in advance' basis.

As mentioned previously, where the construction management division of a building contracting firm is acting as the construction or project manager there is often not much difference between this and the more traditional 'cost plus fixed fee' contract. However, it is more usual for the cost of site staff and facilities to be included in the fee than is the case with the traditional contract.

Summing-up

There are a variety of client requirements, and many contractual methods of satisfying them. The method of contracting for a particular project or group of projects should be chosen with the individual client's needs in mind, not just 'on the usual basis'. It is, for instance, pointless to prepare a BQ for a client who is concerned not with money but with convenience or time.

We must also remember that the client organisation may not really need a building at all – perhaps it should be changing its distribution methods instead of building a new warehouse. It may therefore be important that it should not go initially to somebody who has a vested interest in putting up buildings.

Further reading

Curtis, B., Ward, S. & Chapman, C. (1991) *Roles, Responsibilities and Risks in Management Contracting*. CIRIA Special Publication 81, London.

Franks, J. (1984) *Building Procurement Systems*. Chartered Institute of Building, Ascot.

The Procurement Guide: A Guide to the Development of an Appropriate Building Procurement Strategy. RICS Books, London.

Turner, A. (1997) *Building Procurement*, 2nd edn. Macmillan, London.

Venmore-Rowland, P., Brandon, P. & Mole, T. (eds) (1991) Proceedings of First National RICS Research Conference. *Investment, Procurement and Performance in Construction*. E & FN Spon, London.

Chapter 21
Real-time Cost Control

Why 'real-time'?

Cost planning and control as described so far have been concerned with planning and monitoring costs during the:

- investigation stage,
- planning stage and
- design stage

and finishes at the point where tenders are received, or a contract entered into. During these stages nothing is irrevocable:

- drawings can be redone,
- the scheme can be reduced in size or its configuration altered,
- the whole thing can be started again from scratch or postponed, or
- the scheme can even be totally abandoned

at what will be (from the point of view of the total project cost) a negligible cost for abortive professional work. Even if the project has got as far as the submission of tenders by contractors, the scheme can still be radically changed or abandoned without any commitment to the tenderers or any need to recompense them for their trouble.

However, while the accurate forecasting of a tender amount and the signing of a contract for that sum (or an amended one where the forecast hasn't quite worked out) is of considerable importance, the matter does not end there. The final agreed cost, after both the project and the various financial negotiations arising from it have been completed, is rarely the same as the original contract amount. It is this final figure which the client actually has to pay and which is the building cost for which the finance will have had to be raised and serviced. On profit-oriented projects in particular it is the figure on which the success or failure of the project will be judged.

In many cases the 'contract sum' may be merely an intermediate stage in arriving at this final cost, and a system of cost planning and control which stops at this halfway point will only be of limited use (and possibly not worth the money which it costs). It is important, therefore, that the cost control process continues to the point of completion and hand-over of the project. This continuation is often termed 'post-contract' cost control for obvious reasons. It differs from design cost planning in many ways, and these will depend upon the exact contractual arrangements, but two differences will be inescapable:

- The interests of a contractor or contractors now form an important addition to the considerations involved.
- Major expenditure is currently being incurred and future options proportionately reduced.

It becomes increasingly difficult to make major alterations of any kind at the stroke of a pen, once drawings and specifications are being translated into an expensive organisation of resources and finished work. Even delay caused by hesitation or reconsideration may involve the client in heavy costs.

It is to emphasise this similarity with real-time computer control systems, and because in some circumstances there may not even be a contract, that we have chosen to use the more appropriate term 'real-time' rather than 'post-contract' to describe this type of control.

The basis of real-time cost control is reporting at regular (often monthly) intervals or on a special occasion when a major decision has to be made. This cost control report will set out the client's likely final cost commitment in some detail and also the cost consequences of any remaining major options. In some instances this report may be made to the architect or other professional project controller, but it is preferable that it should be made direct to the client.

This service should be envisaged when the QS's fee is being negotiated, and should be written into the agreement. Its purpose is to:

- enable the client to budget for the likely expenditure;
- enable the cost effect of any major changes to be seen in the context of the project as a whole;
- enable avoiding action to be taken if the total cost appears to be escalating unduly.

One of the problems attached to this service is the extent to which sums of money should be kept in hidden reserve by:

- the underestimation of savings, and
- the overestimation, or at least the pessimistic evaluation, of additional costs.

There is a natural tendency towards this type of caution. In its most extreme forms this may be taken to the extent of inflating the projections sufficiently to cover without disclosure any mistakes which may be made by the architect or quantity surveyor and which will still leave the client with a pleasant surprise when the final account is inevitably settled at below the forecast amount. Such conduct would be as unprofessional and unhelpful as inflating the quantities in the BQ for similar reasons.

Nevertheless, caution dictates that some allowance be made for inevitable minor additions which are not immediately apparent, or for things going slightly wrong. This particularly applies where final costs will be subject to negotiation or where the full consequences of decisions or events cannot be exactly foreseen. The reasonable judgement of these allowances, avoiding both excessive pessimism or optimism, is one of the major factors in making this a truly professional task. In particular, the quantity surveyor who is successful in this role will develop a feeling for the average project, where not everything is going to go wrong, and for the occasional project where it would be wise to assume that it might.

The problem of information

In carrying out real-time cost control it is essential for the QS always to be fully informed as to what is happening and what is intended. Ensuring this may well be one of the more difficult parts of the operation.

Ideally the QS should be part of the decision-making process, in which case the difficulty should not arise, but this is not usually the situation. However, merely sitting in the office waiting for information to come in is not going to be good enough. Most contracts contain no requirement for communication to and from the contractor to flow through the QS's office. The QS is likely to be informed eventually of everything necessary for the final settlement of accounts, but this information will be far too late to be of any use for real-time cost control purposes.

It must be remembered that cost control requires not just a record of costs incurred to date but also of likely eventual cost commitments arising from:

- current proposals for variations;
- other decisions which have been taken by the design team and/or the client which will create variations and/or cause delay or difficulties in working;
- failure by the design team to meet deadlines for supply of information or for appointing nominated sub-contractors, etc., which will have the same effect.

The QS must, therefore, instigate 'current awareness' procedures, and in particular must:

- insist on seeing immediate copies of all official orders, drawings, and letters;
- attend all site meetings and generally look around the site to see what is going on;
- use this opportunity to find out what verbal instructions might have been given.

It has often been alleged that on many projects the 'official' documentation is merely trying to catch up with the real but informal site communications system. The preparation of interim valuations for payment purposes provides an excellent opportunity to monitor what is actually happening and should always be used for this purpose. The maintenance and updating of real-time cost control records is an ideal application for computer systems, whether they be spreadsheet or specialist control systems.

Although, as shown above, many of the factors in real-time cost control are common to any type of contractual situation, it will be most convenient to look at the detail in the context of each of the different types.

Real-time cost control of lump sum contracts based on BQs

In the real-time cost control system for this type of project the starting point will be the contract sum, and the provisions which the contract makes for amending it. These provisions will usually cover:

- The adjustment of 'provisional' sums of money or quantities of work, which are embodied in the contract, in the light of the actual cost or quantity of work carried out. This type of provision is often made for work such as foundation excavation which cannot be foreseen exactly.
- Payment for variations and for additions to, and omissions from, the work.
- The adjustment of amounts included in the contract for work to be carried out by nominated sub-contractors and others, in the light of actual cost.
- The correction of errors in BQs and other contract documents.
- Adjustment for the effects of inflation (if applicable).
- Adjustment for the effects of new legislation.
- Compensation to the contractor for the cost of delays in the work caused by circumstances for which the client is responsible (or, in more generous contracts, by circumstances for which the contractor is not responsible). These circumstances may include delays caused by the instruction of variations or by delay in the provision of drawings, etc., or where the more generous provisions apply, delay caused by bad weather or industrial action.

In addition to the likely revised contract amount the QS's cost projections should include:

- updated additional amounts for professional fees;
- furnishings (unless it has been agreed to exclude these);
- any other items which the client wishes to consider as part of the total cost.

Adjusting the contingency sum

An item which can give rise to controversy is the adjustment of the provisional contingency sum, which is an amount included in the contract sum by the design team to cover unforeseen expenditure. The controversy centres around whether this sum is at the architect's disposal to cover design development (or more bluntly, mistakes) or is at the disposal of the client to spend on extra items.

Some public authorities hold the second view so strongly as to demand the omission of the contingency sum as the first variation on the contract, so as to effectively take it out of the architect's control (while of course retaining it as a buffer in their own calculations).

Except where the client holds such extreme views it is probably best to show the contingency sum separately in the cost control report, deducting any incidental extras of the appropriate kind and carrying the balance forward to cover further similar extras which can occur at any stage of the project. In any case a provision of some kind must be made for contingencies during the remainder of the project. This sum is sometimes split into two, in order to make separate provision for:

- design development and unforeseen circumstances.
- additional items (i.e. extras).

Variations

A general difficulty which faces the QS is that although the contractor is required by most contracts to give written notice to the architect of circumstances having arisen which will give rise to an extra cost, this notice is usually only required to be given within a 'reasonable time' of the occurrence. In contractual terms this has to be interpreted fairly generously, and may well be too late for cost control purposes. The remarks already made about current awareness therefore apply strongly.

A further difficulty under the British JCT Standard Form of Contract (although not always overseas) is that there is no requirement for the financial effect to be stated or agreed at the same time, and in fact this is positively discouraged both by the contract clauses and by traditional practice.

The QS's estimates, therefore, have to bear in mind that the figures may be subject to negotiation, and if the contractor also suggests a figure at this stage (although not legally bound to do so) the QS will have to guess the point between them at which a deal is likely to be made. For this reason the QS's real-time cost projections on this type of contract should never be shown to the contractor, since the QS may have prudently assumed a higher figure for extras than would at that time be conceded.

Although the contract usually contains a similar provision for notification of reduced or omitted work by the contractor there is obviously less incentive for this to be done, and the duty is easily 'forgotten'. The QS again must be in a position to see that omissions are duly notified and adjusted.

Nominated sub-contracts

The growing tendency for main contractors to 'blame' nominated sub-contractors for delays and thereby escape responsibility has led to a decrease in the once-popular use of nominated sub-contractors, although the system is still employed. The adjustment in respect of an amount included for a nominated sub-contractor's work may be required to be done on several different occasions. The first adjustment will be done when a nominated sub-contractor's tender is accepted for the work; this tender is likely to contain similar types of provision for further adjustment as the main contract does, and similar subsequent changes in cost are likely to occur which will have to be reflected in the cost-control report.

Inflation

On 'fluctuation' contracts the adjustment of the cost control report for inflation is a difficult one, particularly at an early stage of the project. It involves:

- a forecast of rises in building costs over the contract period;
- an estimate of the likely progress of the work in cost terms so that the forecast increases can be applied to an appropriate proportion of the cost.

There are three different figures to be calculated for inflation, and this will apply whether the contract adjustment is on the basis of the difference between ascertained actual costs of resources and costs at the time of tender, or whether it is done on an index-linked basis:

- increased cost incurred to date;
- estimated additional cost for remainder of contract of increases announced to date;
- allowances for effect of future increases.

The figure which the client will require for budgeting purposes is the total of the three, although the total should be split into the three categories to emphasise the progressively more conjectural nature of the estimates. Although it is theoretically possible for there to be a decrease in cost through the operation of this clause of a contract there have been no known instances of a reduction in the contract sum from this cause in the UK during the 60 years prior to 1999.

Budgetary control vs best deal

A most important point regarding cost control of lump sum contracts is the perennial conflict between tight budgetary control and getting the best deal from market forces. The best budgetary control is achieved if a firm contractual commitment can be obtained at the earliest possible moment – in the case of extra work, for instance, before the order is given to go ahead. Even the British JCT Standard Form of Contract makes some provision for this in its use of the words 'unless otherwise agreed' in its set of procedures for valuing variations. However, as previously mentioned, the contractor is under no obligation to agree a firm figure beforehand, with its attendant risk of underestimation, and may prefer to leave negotiation until the final costs are available.

It is up to the QS to advise the client on the advantages and disadvantages of agreeing extras at an early stage in the light of:

- the client's particular budgetary problems;
- the apparent willingness or otherwise of the contractor to do a reasonable deal on this basis.

Cost reports

As an example of a system of cost reporting, two consecutive monthly cost reports on the early stages of a job are shown in Figs 21.1 and 21.2. In order to illustrate the

OFFICE BLOCK, MIDTOWN

Cost Report No 2 10/06/00   **Contract sum**   £1,910,200

	Omit £	Add £	£
Net omissions			
Partial substitution of 'Conglint' panels for			
Portland stone (A.I.3)		27,050	
* Saving in foundations and piling		7,000	
		34,050	34,050
			1,876,150
Net additions		£	
Decorations and partitions in offices (A.I.1)		44,000	
Carpeting in offices in lieu of wood block flooring		9,100	
		53,100	/ 53,100
			1,929,250

Adjustment of PC sums

	Omit £	Add £	
Heating installation (Midland Heating Co)		7,500	
Electrical installation (Sparks & Co)		975	
* Curtain walling (The Curtain Walling Co)	2,065		
	2,065	8,475	
		2,065	
		6,410	
Profit and attendance		500	
Net addition		6,910	6,910
			1,936,160

Adjustment for fluctuations (inflation)
including sub-contractors

		£	
*(i) Increase on work to date		2,000	
*(ii) Allowance for effect of current			
increases on remainder of work		60,000	
*(iii) Allowance for possible future increases, say		50,000	
		112,000	112,000
Total estimated final cost			£2,048,160

Amount included for contingencies

		£
Contingency sum in contract		30,000
* Sundry minor variations (net extra)		1,000
Balance remaining		29,000

Note * indicates changed or additional item since previous cost report
VAT and professional fees not included in above figures

Fig. 21.1   Real-time cost report, June.

OFFICE BLOCK, MIDTOWN

| **Cost Report No** 3 12/07/00 | | **Contract sum** | £1,910,200 |

Net omissions £

Partial substitution of 'Conglint' panels for Portland stone (A.I.3)		27,050	
* Saving in foundations and piling (revised amount)		6,652	
		33,702	33,702
			1,876,498

Net additions £

Decorations and partitions in offices (A.I.1)		44,000	
Carpeting in offices in lieu of wood block flooring		9,100	
* Extra work to entrance vestibule and staircase (A.I.6)		6,000	
		59,100	59,100
			1,935,598

Adjustment of PC sums	**Omit**	**Add**	
	£	£	
Heating installation (Midland Heating Co)		7,500	
Electrical installation (Sparks & Co)		975	
* Curtain walling (The Curtain Walling Co)	2,065		
* Lifts (Ascenseurs Ltd)	1,995		
	4,060	8,475	
		4,060	
		4,415	
Profit and attendance		450	
Net addition		4,865	4,865
			1,940,463

Adjustment for fluctuations (inflation)
including sub-contractors £

*(i) Increase on work to date		6,500	
*(ii) Allowance for effect of current increases on remainder of work		65,000	
*(iii) Allowance for possible future increases, say		40,000	
		111,500	111,500
Total estimated final cost			£2,051,963

Amount included for contingencies £

Contingency sum in contract			30,000
* Sundry minor variations (net extra)		2,200	
* Likely claim for delay in connection with Ground floor walling layout (site meeting 05.07.00)		12,000	
		14,200	14,200
Balance remaining			15,800

Note * **indicates changed or additional item since previous cost report**
VAT and professional fees not included in above figures

Fig. 21.2 Real-time cost report, July.

difficulties involved a 'fluctuations' contract has been assumed. The references A.I.1, A.I.3, etc., refer to sequentially numbered architect's instructions.

There are many possible ways of setting out such reports, and readers may well be able to devise what they consider to be a better format. What is most important is that any changes since the previous report should be clearly highlighted, either by the use of asterisks and notes, as in the example, or else by simply giving the previous figures as totals. Using the latter system the figures in the second example for 'Net additions' could simply have been shown thus:

Net additions	£
As before	53,100
Extra work to entrance vestibule and staircase (A.I.6)	6,000
	£59,100

Points where other people's tastes might differ include:

- The rounding off of estimated figures (a figure of £18,577 implies a standard of accuracy which £18,600 does not).
- The item of 'sundry minor variations' which could have been shown in detail.

It is the authors' view that this document is prepared to assist budgetary control, not to start a premature witch hunt against the architect by highlighting minor problems which have arisen. Consideration of such matters can usually be better left until the final account, when these problems can be seen in the context of the finished project.

It will be noted that the report takes account of variations ordered, and quotations accepted, for work which is not yet done or even started. The possible claim for delay has been charged against the contingency fund, which will not be able to stand many such demands upon it.

Finally, it must again be emphasised that under normal contractual arrangements the contractor has no duty to participate in any cost control scheme or cash flow control scheme of the client, and there are sound commercial reasons for refusing to do so. Any such requirement, to be binding, must be written into the form of contract with legal advice and cannot simply be stated in the preliminaries of the BQ.

Real-time cost control of negotiated contracts

A negotiated contract provides excellent opportunities for cost control. Indeed, this is one of its principal advantages for which, as usual, some measure of market-force benefit may have to be sacrificed.

From the moment of involvement in the scheme the contractor should be committed by the architect and QS to participate in the cost control process both during the design stage and subsequently. This requirement should be made clear during the early stages of negotiation and duly written into the contractual arrangements, together with a provision for prior negotiation and agreement of extra costs.

Real-time cost control of cost reimbursement contracts

The situation here is quite different. There is no contract sum as a starting-off point and no possibility of contractual commitment to estimates of original or extra costs, so that real-time cost control by the quantity surveyor becomes of fundamental importance.

On the other hand there is no price mechanism to cloud the issue, and thus no valid commercial reason why the contractor's own estimates and costings should not be made fully available to the client's representatives on a so-called 'open-book' basis.

It is preferable that cost control of the project should be carried out by the quantity surveyor cooperatively with the contractor, since two heads will certainly be better than one and full advantage can be taken of the absence of commercial secrecy, which is one of the main benefits of this type of contract.

At any point in the project the QS's cost control report is likely to be based upon:

- the QS's original estimate of cost, which should have incorporated an allowance for inflation;
- the QS's estimate of cost of variations, which also should have incorporated an allowance for inflation;
- any adjustment of estimated cost for actual cost of work completed or expenditure committed, where this is different to what was envisaged;
- any adjustment to the estimated cost of current or future work in the light of this experience.

Towards the end of the project the QS will probably switch to a simpler system involving the ascertained cost of work completed plus the estimated cost of the remaining work required.

The QS's original estimate is likely to be based upon approximate quantities, and so are the QS estimates of the likely cost of variations. However, in both cases, and especially the latter, a 'resource-based' approach may be adopted in cooperation with the contractor.

Even where approximate quantities are used, however, all estimates should be discussed with the contractor before being given to the client. The purpose of this cooperation is twofold:

- to ensure that the QS has not made any false assumptions;
- to enable a positive contribution to be made where the QS thinks that the methods or equipment which the contractor is proposing to use are unnecessarily expensive or inefficient.

It must be remembered that the builder's usual incentive to cost cutting will be absent, since it is the client's money, not the contractor's, which is being spent. In such an event there may have to be a meeting between the architect, QS and contractor to decide what is to be done; this could in some instances involve minor redesign to enable something to be built more efficiently. This aspect of cost control is an important part of the QS's contribution.

A most difficult part of the whole task will be the replacement of estimated costs by actual ascertained costs. Under some forms used for cost-reimbursement contracts the contractor's responsibility for costs is limited to providing a periodic statement of total labour and plant costs to date, and a list of invoices received. Such a document is totally useless for cost control purposes both in format and in timing, since invoices are often not available until weeks or even months after the cost commitment has been incurred. The QS must, therefore, ensure that the contractual arrangements provide for:

- Facilities for the client's representatives to participate in the contractor's estimating and work planning procedures and in the appointment and control of sub-contractors.
- The use of the contractor's own detailed costing system for the benefit of the client's representatives, or alternatively the installation of a suitable system where the contractors' system is unsuitable or non-existent.
- Early disclosure of all papers concerned with ordering, costing, delivery or pricing.

It must be made clear prior to the contractors' appointment that these requirements exist, and they must be incorporated into the contract. The arrangements must not rely upon goodwill, since at the least they are inconvenient to the contractor and are certain to involve some expense. This will need to be reflected in the contractor's fee.

The costing system will have to separate the costs in respect of specific parts of the work which correspond to identifiable portions of the QS's estimate, otherwise any form of reconciliation between costs and estimate will be impossible until the whole job is complete. This means in turn that the QS's estimate will need to be structured suitably with this need in mind, an elemental basis being quite a good one. However, work executed by different trades at different times should not be telescoped into a single priced item (for example roof carcassing and roof tiling).

A further point to watch is delay in reporting; costs of plant and materials may take some time to work through the contractors' office system and a method has to be devised for overcoming this delay. For example, materials could be priced from delivery dockets – which should be instantly available. Reconciliation of delivered materials with measurements of major items of work should also be carried out to ensure that there are no unexplained differences up or down due to theft, excessive wastage or clerical error.

Reconciliation of allowances for inflation poses some further problems, since these will often not be separately identified in the ascertained costs of resources. In most cases some form of approximation will be adequate in showing how much of the cost of some particular operation should be set against the allowance for inflation, since expensive and complicated book-keeping arrangements which do not affect the total to be paid are unlikely to yield any commensurate benefit to the client. Under this type of contract the client is always going to have to meet the cost of inflation. Allowances for future inflation may need to be revised from time to time, as was the case with cost control of 'fluctuating' lump sum contracts.

Cost control of sub-contractors on cost reimbursement contracts

It is necessary to watch closely the arrangements for sub-contractors. The builder, in collaboration with the QS, will be seeking separate tenders from all the specialist sub-contractors, and each of these has to be dealt with separately. It might seem a good idea to try and obtain the most important tenders at more or less the same time, since it would then be possible to consider their total effect on the cost of the project. However, this is rarely advantageous in practice since cost reimbursement is most likely to be used on projects where design and construction proceed more or less simultaneously, with the design of later stages of the work going on while earlier stages are being executed.

Specialist work should not be tendered for until the design work on that particular specialism has been completed, otherwise unnecessary variations are likely (and this will be no help whatever to cost control). So it is more usual to carry out the cost control of such projects with the budgets for the specialist works contractors being turned into firm prices one by one. This means that it may often be possible to appoint the specialists on a price basis rather than a cost basis, as the work may be more clearly defined by the time they are involved. For example, the structural items to which finishes are to be applied may actually have been built.

Competitive tendering for sub-contracts might well be possible; again if the QS does not look after this then nobody else will. Without such control it is much easier for the builder just to ring up a friendly sub-contractor and make the necessary arrangements.

Leaving the calling of tenders for specialist work until they are needed also has the advantage that it is possible to amend the budgets (and specifications) for later specialists if there are cost over-runs on the earlier ones, so that tenders can be sought on the revised basis. This gives better cost control than having to amend works contracts to obtain savings, as may happen if tenders have been called too early.

Difficulties with combining price-based work and cost reimbursed work on the one project

Great care is necessary when work on a price basis and work on a cost reimbursement basis are both included in the same contract. This is only likely to be satisfactory where one or more of the following circumstances exists:

- The cost reimbursement work is only a negligible proportion of the whole (e.g. incidental dayworks on a lump sum contract).
- The two types of payment apply to two separate organisations (e.g. the main contractor and a sub-contractor).

- The cost reimbursement work is carried out by a separate gang of people using special materials.
- The two types of work are carried out at different periods or at different sites.

Unless at least one of these conditions is fulfilled there will be a tendency for labour and/or materials used in the price-based work to be charged to the cost reimbursement work, thus being paid for twice. This is almost impossible to check without continuous monitoring. An example of the type of project where this could happen would be an extension to an existing building where the extension was to be paid for as a lump sum but the alterations in the existing building were done on a cost reimbursement basis. It would be very difficult to ensure that operatives' time was charged to the correct part of the work if it was all proceeding simultaneously, and even more difficult to check afterwards that this had been done. Such hybrid arrangements should therefore be avoided.

It will have been seen that the real-time cost control of cost reimbursement contracts involves the skilful combination of contractors' costing systems and client budgeting and cost strategies, and is likely to be expensive. There is no halfway house, however; cost control of this type of project can only be done properly or not at all. Clients must make up their minds about this at the beginning, under appropriate advisement. The internal cost control of a package deal contract would have many features in common with the above procedures.

Real-time cost control of management contracts

In many ways the procedures would resemble those on a cost reimbursement contract, but probably at a more strategic level since the work will be undertaken using a number of different contracts, any of which in turn may require its own cost control procedures of the appropriate type already discussed. The remarks made above concerning sub-contracts on cost reimbursement work will apply to all the works contracts.

On a really large project it is not necessary that the whole of a particular work package must be let at once, and there would be advantages in letting a package in several parts at different times if only some areas of the project have been fully designed.

Nevertheless, the general pattern set out on pages 309–14 will also apply to the project as a whole.

A spreadsheet cost report on a management contract

In Fig. 21.3 is shown a financial report (No 5) on a sports training centre. Where reports such as this are being generated by computer they are likely to be updated as changes occur, rather than on a monthly or other period basis.

Spreadsheet output should always be checked arithmetically, as a matter of good practice. Although computers do not make mistakes in carrying out the calculations which they have been instructed to do, it is always possible to give them the wrong instructions in compiling a spreadsheet.

Trade package	Package reference	Estimated value (cost plan) (A)	Transfers within cost plan	Adjusted cost plan	Package sum (B)	Confirmed instructions (C)	Anticipated instructions (D)	Anticipated final cost (B+C+D)	Diff. with adjusted cost plan	Package procurement status
Sub-structure/ground works	1100	282,000.00	81,313.00	363,313.00	370,086.07	3,200.00	5,300.00	378,586.07	15,273.07	Contract let and on site
Steelwork	1200	275,000.00		275,000.00	258,921.00	1,000.00	15,000.00	274,921.00	(79.00)	Contract let and on site
Precast floors and stairs	1300	44,500.00		44,500.00	42,194.00			42,194.00	(2,306.00)	Contract let and on site
Roof system and rainwater goods	1400	199,000.00		199,000.00	192,545.00			192,545.00	(6,455.00)	Contract let
Rooflights, patent glazing, windows, and curtain walling	1500	110,000.00		110,000.00	110,000.00			110,000.00	0.00	Out to tender
Masonry	1600	109,000.00	13,600.00	122,600.00	122,600.00			122,600.00	0.00	Out to tender
Internal partitions	1700	35,000.00		35,000.00	35,000.00			35,000.00	0.00	Out to tender
Joinery, doors and ironmongery	2000	100,000.00	12,000.00	112,000.00	112,000.00			112,000.00	0.00	
Plastering/screeds	2100	60,000.00		60,000.00	60,000.00			60,000.00	0.00	
Decorations/painting	2200	55,000.00		55,000.00	55,000.00			55,000.00	0.00	
Floor finishes	2500	145,000.00		145,000.00	145,000.00			145,000.00	0.00	
Ceilings	2700	45,000.00		45,000.00	45,000.00			45,000.00	0.00	
Mechanical and electrical installations	3000	800,000.00	(81,313.00)	718,687.00	718,687.00			718,687.00	0.00	Out to tender
Builderswork in connection with services	3500	30,000.00		30,000.00	30,000.00			30,000.00	0.00	
Toilet cubicles, back panels and vanity units	4000	40,000.00		40,000.00	40,000.00			40,000.00	0.00	
Loose fittings, furniture and equipment	5000	270,000.00		270,000.00	270,000.00			270,000.00	0.00	
External works and landscaping	6000	220,000.00		220,000.00	220,000.00			220,000.00	0.00	
Sub-total		2,819,500.00	25,600.00	2,845,100.00	2,827,033.07	4,200.00	20,300.00	2,851,533.07	6,433.07	
Contractor's percentage for overheads and profit @ 4.00%		112,780.00	1,024.00	113,804.00	113,081.32	168.00	812.00	114,061.32	257.32	
Contractors preliminaries		291,556.16	—	291,556.16	291,556.16	—	20,000.00	311,556.16	20,000.00	
Package interface (contingencies)		189,900.84	(26,624.00)	163,276.84	182,066.45	—	(45,480.00)	136,586.45	(26,690.39)	
Total project cost (£)		3,413,737.00	0.00	3,413,737.00	3,413,737.00	4,368.00	(4,368.00)	3,413,737.00	0.00	

Fig. 21.3 Bloggsville United Training Centre Financial Summary No 5.

Some points to note in this example:

- The figures exclude VAT and professional fees.

- The 'Estimated value' column represents the cost plan, against which the client's budget has been set. The client may be holding additional project contingencies, not displayed here, as on the 'open book' basis the contractor has input to, and receives, this summary.

- Figures in the 'Transfers within cost plan' column indicate where amounts have been transferred between packages in the cost plan, for example the transfer of drainage from 'M&E Installations' to the 'Groundworks' package. These changes result in the adjusted cost plan, which should always total the same as the 'cost plan', although constituent parts will vary.

- In the 'Package sum (B)' column the figures are the same as in the adjusted cost plan unless tenders have been received and sub-contractors appointed, when the revised figures are inserted (in italics, to make clear that this has been done).

- Columns C and D represent variations to the package values, whether formally instructed or merely anticipated.

- The most important column is the 'Anticipated final cost' (B + C + D) column, which indicates where the project is heading financially, both on a package-by-package basis and overall. It will be seen that this total is at the moment the same as the 'Adjusted cost plan', but note that this has been achieved by a reduction in the 'Package interface (contingencies)' figure. A 'zero' (0.00) in the 'Diff. with adjusted cost plan' column confirms that the anticipated final cost does not exceed the budget.

- The main contractor's (construction manager's) costs and the preliminaries costs are included as a percentage of the package costs. (These were determined by competitive tender.)

- Note that a sum of £200,000 is included for estimated preliminaries costs associated with an anticipated 2-week extension of time relating to problems encountered with the sub-structure and foundations.

Control of cash flow

So far in this chapter we have ignored the phasing of expenditure, except in so far as the effect of inflation on costs is concerned. However, in practice client organisations are usually very concerned about the sums of money they will be called upon to pay at any given time, and may require a cash flow estimate from the quantity surveyor before signing the contract, rather of the type shown in Table 3 (Chapter 6) in respect of the block of flats. They will need this in order to make adequate arrangements to raise the money, irrespective of whether they are operating in the profit sector or in the social sector. Quite clearly they are unlikely to be able to place the whole contract amount in their bank trading accounts at the beginning of the project, ready to be drawn upon as required over the contract period. Even if they were theoretically in a position to do this they would still have more profitable uses for the money in the interim, which would probably mean that it would not be instantly available.

The cash flow estimate will require to be revised by the QS from time to time, preferably at the same intervals as the real-time cost control reports. These revisions will take into account not only any changes in total cost but also the extent to which some aspects of the programme are running faster or slower than anticipated, or indeed whether the whole programme has slipped. In many cases, a periodic report of this kind may be all that is required, giving the client an updated warning of cash requirements. As well as performing this function such a report may also be very useful in monitoring programme slippage, because the level of cash flow required to complete on time may be seen to be quite impossible in the light of experience to date.

However, in some instances a client will need not merely a report of what is likely to happen but some positive form of control. In the case of a cost reimbursement contract it should be possible to order an increased or decreased tempo of work, and few problems arise. In management projects, also, the placing of contracts can be deferred or work accelerated, but serious difficulties occur in the case of lump sum BQ contracts on the Standard Form. Once a contract has been signed using the British JCT Standard Form of Contract the contractor is not answerable for the phasing of expenditure. The contracting firm does not have to provide an estimate of client's cash flow, and even if they agree to do so as a favour they cannot be held to perform in accordance with it. Their only duty is to proceed 'regularly and diligently' and to complete by the appointed date or a later date which may be agreed upon because of delays.

It is in the contractor's interest to obtain high levels of payment in the early stages of the project in order to finance the later stages and so the prices for earlier parts of the work may be inflated, and early deliveries of materials arranged, to achieve this. Therefore if a positive form of cash flow control is required, a suitable provision will have to be made in the actual contract (with legal advice), and not merely in the BQ or in correspondence. The contract might state a month-by-month cash flow programme with provisions that:

- The contractor would not be reimbursed ahead of this programme if the work was carried out faster than scheduled.
- If expenditure fell behind the programme there could either be a provision for damages to be ascertained or, more constructively, for the difference between the programmed amount and the actual expenditure to be paid into a trust fund. It might be thought that there would be no damage to the client's interests if expenditure fell behind the programme (provided the job was finally completed on time), but in the public sector and in some large private corporations construction is often funded on an annual basis and amounts unspent at the end of the financial year may be permanently lost to the authority or department concerned.

In the profit sector, however, quite apart from their general wish to reduce the cost of financing their cash flow, clients may find it difficult to raise the total cost of construction until nearer the time when the building will actually be producing an income. They might therefore arrange for the contractor to bear part of the building cost during construction or even to take a share in the risk of the development. In such cases it would be essential for the arrangements for funding during the progress of the works to be formalised in the contract.

Cash flow control of major development schemes

Major development schemes may extend over many years and may include a large number of building contracts entered into at different times. Cash flow control is necessary in the running of public development schemes to ensure that money is available to meet outgoings. If a development scheme is funded on an accrual basis (that is, unspent funds which have been allocated to it can be carried forward from year to year), and if it is not possible to lend out money at a profit (perhaps because the authority has not got the necessary powers), then nothing more than a 'passive' control of cash flow will be required. The authority will simply need to order its financial affairs to suit what is happening.

However, public funding is often on a yearly basis, in which unspent funds in one sector are not carried forward for the benefit of the project or authority for which they were allocated, but are used to meet overspending in other sectors or applied to a reduction in government expenditure. In addition, most major building projects involve an unavoidable commitment to expenditure for some years ahead, although the authority's budget for these years may be unconfirmed when the commitment is entered into.

Trouble may then occur if major construction projects fall behind schedule, so that much of the expenditure which should have taken place in the current year becomes a commitment in future years, and the cash balance which is left at the end of the current year will be lost to the authority with no certainty of picking it up again when it is needed. It is, therefore, necessary to manipulate cashflow in the light of changes so that:

- The year's allocation of funds is completely but productively spent and a realistic revision can be made of cash requirements for ensuing years.
- The ongoing programme can be hastily amended if funds are reduced (or increased!) in any particular year.

This could be called 'active' or 'positive' control of cash flow.

In any authority's development budget the expenditure will fall into three categories, and in each of these categories the discretionary factor in expenditure is progressively increased.

Old projects. These are projects which have been completed but for which full payment has not yet been made (possibly depending upon the outcome of litigation, arbitration, or negotiation). Once payments are due they will have to be met.

Current projects. It may be possible, and worthwhile, for an authority to order a speed-up or slow-down of current projects in order to change its cash commitment during the year, even at the expense of an eventual increase in total cost. What more usually happens, however, is that circumstances beyond the authority's control cause such changes, and the cash flow budgeting has to be revised to suit. Almost invariably, expenditure tends to fall behind estimate because of delays in progress, although this may sometimes be counterbalanced to some extent by extras and by inflation.

It might be thought that if an authority had a large number of projects on hand (such as a programme of 10 schools) the differences would tend to average out. But this tends not to happen in practice because the causes of delay are often national or regional in character – weather, industrial disputes, labour shortages, etc.

New projects. It is these which offer the greatest scope for control. If the start of a 2-year project worth £10,000,000 is brought forward or put back by one month the effect is likely to be an increase or decrease of £400,000 in the current year's cash flow, and correspondingly greater for longer periods than one month. There are two major difficulties however:

- The starting of a major project may have a comparatively small effect in the year in which the go-ahead is given (but it will then join the ranks of the current projects and will be an inescapable and major commitment for the following 2 years or more).
- Expenditure on projects tends to follow the well-known 'S' curve where the rate of spending is at its greatest in the middle period of the project, with a comparatively slow build-up in the early stages.

In many ways the best way to control the cash flow situation is by having a number of smaller short-term projects ready to roll. These can be largely completed during the financial year in question, even if the order is not given until the year is quite well advanced.

An obvious method of disbursing funds on a project which is falling behind programme is to pay for the work, or for materials, ahead of the legal liability to do so. There are obvious dangers in doing this and a public auditor would be unlikely to accept a straightforward attempt to do it. It might, however, be possible to pay money into a joint trust fund.

Control of short-term cash flow

Up to now we have been considering cash flow as though we are simply concerned with the year as a whole, but this is rarely the case. Funds are not normally paid across to an authority in a single sum at the beginning of the year, nor are they paid as each cash commitment arises, but are usually paid in quarterly instalments or something of the kind.

Quite clearly, therefore, the authority must avoid too much cash expenditure in the early part of the year since funds will not be available, whilst if it is empowered to lend surplus short-term money profitably it may be able to arrange matters so that it can do this. With quarterly funding, even where there is a smooth monthly outgoing throughout the year, the money for the third month of any quarter will be available for alternative use during the first 2 months.

Payment delays on profit projects will be to the client's advantage

Although the above section has been written in the context of public or social development, much of it is equally applicable to profit schemes, except that the problem of 'spending all the money by the end of the year' is not so likely to occur. Any pressure is likely to be in the opposite direction; the question of interest on money borrowed becomes of paramount importance so that any pushing back of payment dates will be welcomed, so long as final completion is not held up. It is not unknown for this to happen in the public domain.

An overseas regional government, which was anxious to minimise the effects of a slump in the local building industry, but had spent all its funds for the year, let contracts on the basis that no payment would be made until the following financial year, so that for the first 6 months or more the contractors were totally financing the programme. However, as would be expected, the tenders reflected this additional burden.

Cost control on a resource basis

As with real-time cost control on any other basis this fundamentally depends upon swift and reliable reporting of what is actually happening, and the rapid reconciliation of this with the control document.

The cost controller will be concerned with two different aspects. The first is the overall time for the project. The preparation of the network will have involved a large number of assumptions about the time to be taken by the various activities, and revisions must be made in the light of actual progress. In this context it is only the activities on the critical or near-critical paths which have to be considered, but any substantial change (for better or worse) may cause a change in the critical path itself and in the duration of the works. Decisions will then have to be made as to whether a different duration is acceptable, or whether the time for remaining activities must be adjusted somehow. In both cases cost will be substantially affected.

The second aspect is the cost of individual activities. A costing system has three objectives:

- to measure actual expenditure against estimated cost;
- to indicate whether performance needs to be improved;
- to provide feedback which will improve future estimating performance.

Costs are recorded under 'cost centres'. These can be anything from the whole project to a tiny part of it, but to meet the above objectives costs need to be recorded under headings which can be identified with sections of the estimate. The estimate in turn needs to be prepared in a suitable format to achieve this.

There is a need to establish a level of breakdown which is neither so fine as to make the recording and allocation of labour and material costs unduly detailed, nor so crude that the 'cost centres' are too broad to mean anything. Individual BQ items are normally at too fine a level, a whole trade section would be too crude. A single activity, or a group of activities, from the network is usually a suitable basis.

Costing of site work has two main difficulties which distinguish it from normal production costing as practised in factories and as described in most textbooks on costing. Both difficulties spring from the one-off nature of most of the work in terms of the finished product and of the conditions under which it is produced.

The first problem is that the usual costing system of 'standard costs' does not really apply. A standard cost is an ascertained cost of carrying out an operation which can then be built into an estimating system, and which should normally be achieved in future – any substantial deviation requiring to be investigated. Machining operations in a wood-working shop can be costed on this basis – once a standard sequence of events has been carried out on a machine a number of times (inserting the work, operating on it, withdrawing it) there is no reason why the same operation should not always cost exactly the same in the future.

Site operations are not only less standardised than this, but the position of the work, weather conditions, etc., etc., varies continuously. It is very difficult to get a sufficient run of identical work to establish standard costs and still have enough similar work ahead to make it worthwhile to apply them. Certainly it is not usually possible to apply such data from job to job, as a joinery works or engineering works would do.

The second difficulty is that of actually recording and processing costs of labour and materials. When operatives are dispersed about the site it is almost impossible to record accurately what they are doing except in fairly general terms, unless a quite unacceptably expensive staff is engaged for the purpose. Many sophisticated costing systems have failed through the time sheets being filled in on a Friday afternoon in the foreman's office from memory.

Materials to some extent are an equal problem, as nothing like a factory control system is operated on most sites when materials are drawn from stock. The invoice is often the main weapon in materials cost control, so it is convenient if the cost centres are sufficiently large and identifiable for particular invoices to be identified with them. However, some contractors' systems do not bother overmuch about materials. As we have already seen, these are much less liable than labour and plant costs to deviate from estimate, given a reasonable level of management. This is another advantage, from the contractor's point of view, of not using 'all-in' labour-and-material rates.

All this may sound very defeatist – difficulties exist to be overcome, not to be excused. They can indeed be overcome, but except in some limited circumstances (such as repetitive house building) nobody has found that the cost and trouble of doing this have produced economically worthwhile results, because of the already mentioned difficulty of deriving 'standard costs' from the data for re-use. All one is likely to get is an expensive post-mortem report. A further point to remember is that some of the people on a site, particularly labourers, are doing work which has to be done but which may not have been identified as an operation in the network – sweeping up, carrying odd materials, attending on the Clerk of Works.

It is, therefore, preferable to use fairly broad cost centres 'fixing formwork to second storey', and devote one's attention to getting the feedback rapidly and with a fair degree of reliability. It should then be possible to measure actual expenditure against estimated cost with reasonable accuracy, and also to a somewhat lesser extent fulfil the other two objectives of a costing system.

In the reconciliation of costs and estimate, the difficulties concerning inflation, and alterations to the scope of the work, which were mentioned in Chapter 19 will again be encountered. However, the inflation problem can be minimised by working in operative-hours and machine-hours rather than money, for comparison purposes. A contractor's costing system will in any case need to highlight the cost of alterations, because very often these will form an extra to be charged to the client. It will also be necessary to ensure that the cost performance of sub-contractors is kept up-to-date in order to arrive at the projected total for the project, complicated by the fact that these may often be on a price, rather than a cost reimbursement, basis.

Summing-up

'Real-time' or 'post-contract' cost control differs from design cost planning in many ways, and these will depend upon the exact contractual arrangements, but two differences will be inescapable:

- The interests of a contractor or contractors now form an important addition to the considerations involved.
- Major expenditure is currently being incurred and future options proportionately reduced.

However today most clients are even more interested in the final figure than in the amount of the tender. Real-time cost control depends above all things on the QS/cost planner being kept informed as to what is going on. If necessary requirements to this end must be included in the contract arrangements.

Regular reports on the financial progress of the job must be submitted to the client during the progress of the work, so that any necessary steps to prevent cost over-runs can be taken before it is too late.

Chapter 22
Cost Planning and Control of Refurbishment and Repair Work

Thirty or more years ago it was customary to think of building projects in terms of 'green field' projects, that is to say projects erected on an open site (preferably one which had never been built on previously). The classical cost planning techniques which grew up in this era tended to be oriented towards such projects as being the norm.

This was rather a convenient assumption, as it avoided a lot of complications, but it is doubtful whether it was ever completely true. However, today it is clearly untrue, and a very substantial part of the current UK construction programme comprises:

- refurbishment;
- renewal work;
- major repair work.

As the last two categories nearly always include some element of improvement, and as refurbishment will usually involve renewal and repair, they can all be considered together for the purposes of cost planning.

Some of what is written here will also even apply to quite a lot of the so-called 'new' work which is being undertaken and which involves building on very restricted urban sites between existing buildings with difficult access and close proximity to the public.

Lump sum competitive tenders inappropriate

The uncertainties of refurbishment work mean that it will usually be impossible, and certainly inadvisable, to undertake the project on the traditional basis of lump sum competitive tenders for the whole of the works. This is because such a basis:

- requires that the work to be done can be accurately foreseen, and
- formalises a confrontational relationship between contractor and client's representatives which is quite wrong for this type of work.

There are exceptions to this, of course, but not very many. The major exception is where a series of similar buildings are being refurbished (on local authority housing estates for example), and experience with the first few projects by both design team and contractors will enable a schedule of rates, or even a BQ, to be prepared and priced. But normally other more collaborative methods of procurement will usually have to be used – either cost-plus or some form of management contracting.

Elemental cost planning inappropriate

Although it is obviously possible to set out any estimates in an elemental format, the normal process of elemental cost planning is not really appropriate. For one thing it will not be practicable to make the element-by-element cost comparisons with other projects which lie at the heart of this technique. The costs will depend on the state of the individual building in relation to what is proposed to be done with it, and comparisons with other projects on an elemental basis would be meaningless.

So the cost planning and control of refurbishment work needs to start from first principles.

Conflict of objectives

The most important initial step is to recognise the conflict of objectives which is inherent in refurbishment work. There is of course a conflict between cost, time, quality and size in carrying out straightforward new work, but on such projects the client's interest is simply to optimise the scheme in these terms, and elemental cost planning allows this to be done. However, in refurbishment work these basic objectives are usually supplemented, or even outweighed, by major secondary objectives.

Occasionally a single objective, such as speed of completion, may be strongly dominant and this is little different to new work. But more often two or more conflicting objectives are perceived as dominant – for instance, continuing occupation required by a 'user' client may hinder the achievement of speedy completion which is the priority of the 'owner' client.

Even the emergence of a single dominant secondary objective, such as safety of the structure or of the public, may have consequences which cut across normal procedures, particularly procedures for efficient construction methods or financial control.

The likely extent of conflict can only be identified after a thorough assessment of risks and objectives, and the trade-offs between them. Frequent reviews of the relative benefits compared with the likely cost, time, and quality implications will probably be required.

Uncertainty

Uncertainty is a major characteristic of refurbishment work. Generally there is a high level of uncertainty, not only in client objectives but also in available physical data. Such uncertainties will probably extend into the construction period with a high likelihood that unforeseen events will occur.

Problems due to uncertainty may be mitigated by:

- allowing a longer lead-in time;
- assembling the work into discrete packages;
- bringing the contractor (or construction manager) into the team at an early stage.

Even so there will remain a need for client, designer and constructor to respond quickly to the discovery of previously unknown features of, or defects in, the existing building.

Safety

Safety is perhaps the most dominant feature of refurbishment work, and the problems in this area are intensified by the general uncertainty already referred to. Safety affects:

- operatives
- the users of the building and
- the general public

and may relate either to the safety of the structure or to the safety of the operations.

Clearly where safety and cost are in conflict then safety must be paramount, so that it is vital that safety issues be identified at an early stage if estimates are to be relied upon. However, safety is not discretionary so that there are no real choices to be made as regards the level of safety, only the means of ensuring it.

Occupation and/or relocation costs

Another major issue does, however, involve fundamental choice, and that is the extent to which (if at all) the building should remain in use during the works. Here we can be dogmatic. Just as real estate agents are supposed to believe that the three most important things about a property are location, location and location, so the three most important decisions in connection with refurbishment are undoubtedly occupation, occupation and occupation!

The effects on costs, programme and safety if the building remains in use are so enormous that a client who wishes to do this should be encouraged to think again – and again – to see if some other alternative can be found. It is incumbent on the cost planner to keep pointing this out. A problem in this regard is that, seen in advance, the difficulties of moving out seem much greater than the difficulties of staying put.

Staying put surely only means having to put up with a little dirt, noise and inconvenience for a year or so, whereas moving out will involve:

- finding alternative premises;
- moving in to them, and out again at completion;
- getting customers and staff to come to terms with the new location;
- reprinting stationery and literature;

and so on.

Seen in advance, moving out appears to be far more of a hassle. This is an illusion!! Staying put involves much more than slight inconvenience!

A few years ago the railway authorities found it worthwhile to close Crewe station completely for several weeks, and divert all trains to other routes, in order to rearrange the tracks and signalling. The job could certainly have been done piecemeal without such drastic action but the time, cost and safety aspects would all have been badly affected, and there would have been just as much inconvenience, only spread over a longer time scale. And exactly the same situation applies to buildings.

If relocation is decided upon it may be worthwhile to make the timing of the whole project dependent upon the availability of suitable alternative accommodation, rather than deciding on the programme first and then trying to see if anything is available at that time.

If relocation does prove to be impossible or unacceptable then a very realistic allowance must be made for the additional cost and time involved. However, this will not represent the total penalty imposed by the decision – the wear and tear on staff and annoyance to customers or residents, caused by dust and dirt, continuous noise of jack-hammers, temporary access arrangements, etc., will not appear on the project balance sheet but will show itself in staffing difficulties and reduced patronage.

Even if most of the work is carried out at weekends and at night the dust problem is difficult to overcome. Finding stock, papers, furniture, etc., covered in dust every morning is one of the greatest irritants to occupants and tends to lead to complaints about everything else. These problems are bad enough for workers and casual users in public and commercial buildings, but are even worse if peoples' homes are concerned.

Unless it is possible to move them out, occupiers will be subjected to the effects of noise, vibration and dust on a more-or-less permanent basis, and the option of mitigating the nuisance by working at night or at week-ends will not of course be available. Householders are likely to be a good deal more militant than office or shop workers when subjected to these annoyances, and the presence of children on the site will give rise to problems of their safety on the one hand and of security of the works and vandalism on the other.

The need for a liaison manager

The problems connected with refurbishment of buildings in occupation underline the importance of the client having a very senior manager employed full-time to liaise between the building team and the occupants, with ultimate authority to over-ride either party. The client's buildings officer, who is likely to be the first person thought of for this role, is unlikely to be suitable. In the case of the refurbishment of commercial or public buildings this person will not be senior enough to be able to dictate to departmental managers, and is probably more used to being ordered around by them. And in the case of housing refurbishment the role of liaison between residents and contractors will be a particularly demanding one, requiring a person with exceptional interpersonal skills.

This cost of the time of a senior person – who in a commercial organisation would need to be at Board level – must be allowed for. Allowance will also have to be made for the cost consequences of the decisions which this liaison manager will have to make in order to keep the client's organisation running as smoothly as possible or in order to satisfy the reasonable demands of tenants.

It is almost impossible, within the bounds of reason, to overestimate the total costs of refurbishing a building which is in occupation!

Costs excluding occupation

Even without occupation the costs of refurbishment are difficult to estimate, because of the factors to do with uncertainty which have already been mentioned.

But before any figures are given the position of the building with regard to 'listed status' must be investigated, since the planning authorities may be able to impose extremely expensive requirements on the scheme. The cost of obtaining, and working with, obsolete or obsolescent materials if this is either required or dictated is likely to be considerable.

A further point to notice is that the cost of temporary works – scaffolding, shoring, etc., is almost certain to play a much greater part than in the case of new works. Where a major gutting of a building's interior is taking place within retained external walls such works are likely to be very sophisticated and expensive indeed. This is a type of work where the average cost planner's knowledge of both technology and cost is sometimes weaker than it is with new permanent works. It is therefore essential that a structural engineer be involved in the estimates at an early stage if any major structural work is contemplated.

The advantages of working with firms who have a good track record of refurbishment work cannot be overestimated, and this applies equally to design and management consultants as it does to construction staff and works contractors. Again, however, such firms are unlikely to be at the cheapest end of the range.

This really leads to the conclusion that if there is a choice between new-build and refurbishment for the client's premises the problems must be faced squarely at the start – tight, overoptimistic estimating has absolutely no place here. Nobody likes cost overruns at any time, but if they mean that with hindsight the wrong decision has been taken by the client (with professional advice) then the consequences are likely to be more than usually serious.

The cost control process should follow the usual real-time methods set out in Chapter 21. A particular point to watch, however, concerns the letting of work packages to trade or specialist contractors. On new-build work it is always considered advantageous to firm these up fairly early in the project, so that actual quotations can go into the estimated cost in place of the cost planner's estimates. However, it is often better in refurbishment work to wait until the efforts of other trades have reached a point where the specialist's work package can be properly defined and a firm price obtained (perhaps even in competition). In the case of a large project the inconvenience of letting, say, the plastering work in several separate packages could well be justified by the ability to get firm prices for each stage of the work.

Finally, it hardly needs to be said that being aware, and keeping the client aware, of the current cost situation is even more important than usual on this type of project, where in spite of everybody's efforts the cost commitment is always liable to escalate at fairly short notice.

Summing-up

Uncertainty is a major characteristic of refurbishment work, and this means that it will usually be inadvisable to undertake the project on the traditional basis of lump sum competitive tenders for the whole of the works. So other more collaborative methods of procurement have to be used – either cost-plus or some form of management contracting. And although it is obviously possible to set out any estimates in an elemental format the normal process of elemental cost planning is not really applicable, and cost planning must start from first principles.

The most important initial step is to recognise the conflict of objectives which is inherent in such work. The effect on costs, programme and safety if the building remains in use are so enormous that a client who wishes to do this should be encouraged to see if some other alternative can be found. If not, it is important for the client to have a very senior manager employed full-time to liaise between the building team and the occupants, with ultimate authority to over-ride either party.

Further Reading

BMI (1997) *Maintenance Cost Study*. RICS Books, London.
CIRIA (1994) *A Guide to the Management of Building Refurbishment*. CIRIA Report 133. CIRIA, London.

Appendix A
List of Elements from Standard Form of Cost Analysis

This appendix contains the element definitions from the Standard Form of Cost Analysis, but not the full requirements for unit quantity costs, etc., for which the Form itself should be consulted. (Note: The amplified form (i.e. to the sub-element level) is no longer published by the BCIS but is provided here because some firms still use it within their offices.)

1   Sub-structure

All work below underside of screed or where no screed exists to underside of lowest floor finish including damp-proof membrane, together with relevant excavations and foundations.

Notes
(1) Where lowest floor construction does not otherwise provide a platform, the flooring surface shall be included with this element (e.g. if joisted floor, floor boarding would be included here).
(2) Stanchions and columns (with relevant casings) shall be included with 'Frame' (2.A).
(3) Cost of piling and driving shall be shown separately stating system , number and average length of pile.
(4) The cost of external enclosing walls to basements shall be included with 'External walls' (2E) and stated separately for each form of construction.

2.A   Frame

Loadbearing framework of concrete, steel, or timber. Main floor and roof beams, ties and roof trusses of framed buildings. Casing to stanchions and beams for structural or protective purposes.

Notes
(1) Structural walls which form an integral part of the loadbearing framework shall be included either with 'External Walls' (2.E) or 'Internal walls and partitions' (2.G) as appropriate.

(2) Beams which form an integral part of a floor or roof which cannot be segregated therefrom shall be included in the appropriate element.

(3) In unframed buildings roof and floor beams shall be included with 'Upper floors' (2.B) or 'Roof structure' (2.C.1) as appropriate.

(4) If the 'Stair structure' (2.D.1) has had to be included in this element it should be noted separately.

2.B   Upper floors

Upper floors, continuous access floors, balconies and structural screeds (access and private balconies each stated separately), suspended floors over or in basements stated separately.

Notes
(1) Where floor construction does not otherwise provide a platform the flooring surface shall be included with this element (e.g. if joisted floor, floor boarding would be included here).

(2) Beams which form an integral part of a floor slab shall be included with this element.

(3) If the 'Stair structure' (2.D.1) has had to be included in this element it should be noted separately.

2.C   Roof

2.C.1   Roof structure

Construction, including eaves and verges, plates and ceiling joists, gable ends, internal walls and chimneys above plate level, parapet walls, and balustrades.

Notes
(1) Beams which form an integral part of a roof shall be included with this element.

(2) Roof housings (e.g. lift motor and plant rooms) shall be broken down into the appropriate constituent elements.

2.C.2   Roof coverings

Roof screeds and finishings. Battening, felt, slating, tiling, and the like.
Flashings and trims.
Insulation.
Eaves and verge treatment.

2.C.3   Roof drainage

Gutters where not integral with roof structure, rainwater heads, and roof outlets.
(Rainwater downpipes to be included in 'Internal drainage' (5.C.1).)

2.C.4   Rooflights

Rooflights, opening gear, frame, kerbs, and glazing.
Pavement lights.

2.D   Stairs

2.D.1   Stair structure

Construction of ramps, stairs, and landings other than at floor levels.
Ladders.
Escape staircases.

Notes
(1)   The cost of external escape staircases shall be shown separately.
(2)   If the staircase structure has had to be included in the elements 'Frame'
       (2.A) or 'Upper floors' (2.B) this should be stated.

2.D.2   Stair finishes

Finishes to treads, risers, landings (other than at floor levels), ramp surfaces,
strings and soffits.

2.D.3   Stair balustrades and handrails

Balustrades and handrails to stairs, landings, and stairwells.

2.E   External walls

External enclosing walls including that to basements but excluding items inclu-
ded with 'Roof structure' (2.C.1).
Chimneys forming part of external walls up to plate level.
Curtain walling, sheeting rails, and cladding.
Vertical tanking.
Insulation.
Applied external finishes.

Notes
(1)   The cost of structural walls which form an integral and important part of the
       loadbearing framework shall be shown separately.
(2)   Basement walls shall be shown separately and the quantity and cost given
       for each form of construction.
(3)   If walls are self-finished on internal face, this shall be stated.

2.F   Windows and external doors

2.F.1 Windows

Sashes, frames, linings, and trims.
Ironmongery and glazing.
Shop fronts.
Lintels, sills, cavity damp-proof courses and work to reveals of openings.

2.F.2 External doors

Doors, fanlights, and sidelights.
Frames, linings, and trims.
Ironmongery and glazing.
Lintels, thresholds, cavity damp-proof courses and work to reveals of openings.

2.G   Internal walls and partitions

Internal walls, partitions and insulation.
Chimneys forming part of internal walls up to plate level.
Screens, borrowed lights and glazing.
Moveable space-dividing partitions.
Internal balustrades excluding items included with 'Stair balustrades and hand-rails' (2.D.3).

Notes
(1)   The cost of structural walls which form an integral and important part of the loadbearing framework shall be shown separately.
(2)   The cost of proprietary partitioning shall be shown separately stating if self-finished. Doors, etc., provided therein together with ironmongery, should be included stating the number of units installed.
(3)   The cost of proprietary WC cubicles shall be shown separately stating the number provided.
(4)   If design is cross-wall construction, the specification shall be stated and the cost shown separately.

2.H   Internal doors

Doors, fanlights, and sidelights.
Sliding and folding doors.
Hatches.
Frames, linings, and trims.
Ironmongery and glazing.
Lintels, thresholds and work to reveals of openings.

3.A   Wall finishes

Preparatory work and finishes to surfaces of walls internally.
Picture, dado and similar rails.

Notes
(1)   Surfaces which are self-finished (e.g. self-finished partitions, fair faced work) shall be included in the appropriate element.
(2)   Insulation which is a wall finishing shall be included here.
(3)   The cost of finishes applied to the inside face of external walls shall be shown separately.

3.B   Floor finishes

Preparatory work, screeds, skirtings, and finishes to floor surfaces excluding items included with 'Stair finishes' (2.D.2), and structural screeds included with 'Upper floors' (2.B).

Note
Where the floor construction does not otherwise provide a platform the flooring surface will be included either in 'Sub-structure' (1) or 'Upper floors' (2.B) as appropriate.

3.C   Ceiling finishes

3.C.1   *Finishes to ceilings*

Preparatory work and finishes to surfaces of soffits excluding items included with 'Stair finishes' (2.D.2) but including sides and soffits of beams not forming part of a wall surface.
Cornice, coves.

3.C.2   *Suspended ceilings*

Construction and finishes of suspended ceilings.

Notes
(1)   Where ceilings principally provide a source of heat, artificial lighting, or ventilation, they shall be included with the appropriate 'Services' element and the cost shall be stated separately.
(2)   The cost of finishes or suspended ceilings to soffits immediately below roofs shall be shown separately.

4.A   Fittings and furnishings

4.A.1   Fittings, fixtures, and furniture

Fixed and loose fittings and furniture including shelving, cupboards, wardrobes, benches, seating, counters, and the like.
Blinds, blind boxes, curtain tracks, and pelmets.
Blackboards, pin-up boards, notice boards, signs, lettering, mirrors, and the like.
Ironmongery.

4.A.2   Soft furnishings

Curtains, loose carpets, or similar soft furnishing materials.

4.A.3   Works of art

Works of art if not included in a finishes element or elsewhere.

Note
Where items in this element have a significant effect on other elements a note should be included in the appropriate element.

4.A.4   Equipment

Non-mechanical and non-electrical equipment related to the function or need of the building (e.g. gymnasia equipment).

5.A   Sanitary appliances

Baths, basins, sinks, etc.
WCs, slop sinks, urinals, and the like.
Toilet-roll holders, towel rails, etc.
Traps, waste fittings, overflows, and taps as appropriate.

5.B   Services equipment

Kitchen, laundry, hospital and dental equipment, and other specialist mechanical and electrical equipment related to the function of the building.

Note
Local incinerators shall be included with 'Refuse disposal'.

5.C   Disposal installations

5.C.1   Internal drainage

Waste pipes to 'Sanitary appliances' (5.A) and 'Services equipment' (5.B).
Soil, anti-syphonage, and ventilation pipes. Rainwater downpipes.
Floor channels and gratings and drains in ground within buildings up to external face of external walls.

Note
Rainwater gutters are included in 'Roof drainage' (2.C.3).

5.C.2   Refuse disposal

Refuse ducts, waste disposal (grinding) units, chutes, and bins.
Local incinerators and flues thereto.
Paper shredders and incinerators.

5.D   Water installations

5.D.1   Mains supply

Incoming water main from external face of external wall at point of entry into building including valves, water meters, rising main to (but excluding) storage tanks, and main taps. Insulation.

5.D.2   Cold water service

Storage tanks, pumps, pressure boosters, distribution pipework to sanitary appliances and to services equipment.
Valves and taps not included with 'Sanitary appliances' (5.A) and/or 'Services equipment' (5.B). Insulation.

Note
Header tanks, cold water supplies, etc., for heating systems should be included in 'Heat source' (5.E).

5.D.3   Hot water service

Hot water and/or mixed water services.
Storage cylinders, pumps, calorifiers, instantaneous water heaters, distribution pipework to sanitary appliances and services equipment. Valves and taps not included with 'Sanitary appliances' (5.A) and/or 'Services equipment' (5.B). Insulation.

5.D.4   Steam and condensate

Steam distribution and condensate return pipework to and from services equipment within the building including all valves, fittings, etc. Insulation.

Note
Steam and condensate pipework installed in connection with space heating or the like shall be included as appropriate with 'Heat source' (5.E) or 'Space heating and air treatment' (5.F).

5.E   Heat source

Boilers, mounting, firing equipment, pressurising equipment instrumentation and control, ID and FD fans, gantries, flues and chimneys, fuel conveyors, and calorifiers. Cold and treated water supplies and tanks, fuel oil and/or gas supplies, storage tanks, etc., pipework (water or steam mains) pumps, valves, and other equipment. Insulation.

Notes
(1)   Chimneys and flues which are an integral part of the structure shall be included with the appropriate structural element.
(2)   Local heat source shall be included with 'Local heating' (5.F.4).
(3)   Where more than one heat source is provided each shall be analysed separately.

5.F   Space heating and air treatment

1   Heating only by

5.F.1   Water and/or steam

Heat emission units (radiators, pipe coils, etc.) valves and fittings, instrumentation and control, and distribution pipework from 'Heat source' (5.E).

5.F.2   Ducted warm air

Ductwork, grilles, fans, filters, etc., instrumentation and control.

5.F.3   Electricity

Cable heating systems, off-peak heating systems, including storage radiators.

Note
Electrically operated heat emission units other than storage radiators should be included under 'Local heating' (5.F.4).

5.F.4   Local heating

Fireplaces (except flues), radiant heaters, small electrical or gas appliances, etc.

5.F.5   Other heating systems

2   Air treatment

Notes
(1)   System described as having:
'Air treated locally' shall be deemed to include all systems where air treatment (heating or cooling) is performed either in or adjacent to the space to be treated. 'Air treated centrally' shall be deemed to include all systems where air treatment (heating or cooling) is performed at a central point and ducted to the space to being treated.
(2)   The combination of treatments used shall be stated, i.e.:

Heating	Dehumidification or drying
Cooling	Filtration
Humidification	Pressurisation

and whether inlet extract or recirculation.
(3)   High velocity system shall be identified as:

Fan coil	Induction units 2 pipe
Dual duct	Induction units 3 pipe
Reheat	Induction units 4 pipe
Multi-zone	Any other system (state which).

5.F.6   Heating with ventilation (air treated locally)

Distribution pipework ducting, grilles, heat emission units including heating calorifiers except those which are part of 'Heat source' (5.E) instrumentation and control.

5.F.7   Heating with ventilation (air treated centrally)

All work as detailed under (5.F.6) for system where air treated centrally.

5.F.8   Heating with cooling (air treated locally)

All work as detailed under (5.F.6) including chilled water systems and/or cold or treated water feeds. The whole of the cost of the cooling plant and distribution pipework to local cooling units shall be shown separately.

5.F.9   Heating with cooling (air treated centrally)

All work detailed under (5.F.8) for system where air treated centrally.

Note
Where more than one system is used, design criteria, specification notes and costs should be given for each.

5.G Ventilating systems

Mechanical ventilating system not incorporating heating or cooling installations including dust and fume extraction and fresh air injection, unit extract fans, rotating ventilators and instrumentation and controls.

5.H Electrical installations

5.H.1 Electric source and mains

All work from external face of building up to and including local distribution boards including main switchgear, main and sub-main cables, control gear, power factor correction equipment, stand-by equipment, earthing, etc.

Notes
(1)   Installations for electric heating ('built-in' systems) shall be included with 'Space heating and air treatment' (5.F.3).
(2)   The cost of stand-by equipment shall be stated separately.

5.H.2 Electric power supplies

All wiring, cables, conduits, switches from local distribution boards, etc, to and including outlet points for the following:
General purpose socket outlets.
Services equipment.
Disposal installations.
Water installations.
Heat source.
Space heating and air treatment.
Gas installation.
Lift and conveyor installations.
Protective installations.
Communication installations.
Special installations.

Note
The cost of the power supply to these installations should, where possible, be shown separately.

5.H.3 Electric lighting

All wiring, cables, conduits, switches, etc., from local distribution boards and fittings to and including outlet points.

5.H.4 Electric lighting fittings

Lighting fittings including fixing.
Where lighting fittings supplied direct by client, this should be stated.

5.I Gas installation

Town and natural gas services from meter or from point of entry where there is no individual meter: distribution pipework to appliances and equipment.

5.J Lift and conveyor installations

5.J.1 Lifts and hoists

The complete installation including gantries, trolleys, blocks, hooks and ropes, downshop leads, pendant controls, and electrical work from and including isolator.

Notes
(1) The cost of special structural work, e.g. lift walls, lift motor rooms, etc., shall be included in the appropriate structural elements.
(2) Remaining electrical work shall be included with 'Electric power supplies' (5.H.2).
(3) The cost of each type of lift or hoist shall be stated separately.

5.J.2 Escalators

As detailed under 5.J.1.

5.J.3 Conveyors

As detailed under 5.J.1.

5.K Protective installations

5.K.1 Sprinkler installations

The complete sprinkler installation and CO_2 extinguishing system including tanks, control mechanism, etc.

Note
Electrical work shall be included with 'Electrical power supplies' (5.H.2).

5.K.2 Fire-fighting installations

Hosereels, hand extinguishers, asbestos blankets, water and sand buckets, foam inlets, dry risers (and wet risers where only serving fire-fighting equipment).

5.K.3 Lightning protection

The complete lightning protection installation from finials conductor tapes, to and including earthing.

Note

The cost of lightning protection to boiler and vent stacks shall be stated separately.

5.L   Communication installations

The following installations shall be included:

Warning installations (fire and theft)
   Burglar and security alarms.
   Fire alarms.
Visual and audio installations
   Door signals.
   Timed signals.
   Call signals.
   Clocks.
   Telephones.
   Public address.
   Radio.
   Television.
   Pneumatic message systems.

Notes
(1)   The cost of each installation shall be stated separately if possible along with an indication of the specification.
(2)   The cost of the work in connection with electrical supply shall be included with 'Electric power supplies' (5.H.2).

5.M   Special installations

All other mechanical and/or electrical installations (separately identifiable) which have not been included elsewhere, e.g. Chemical gases; Medical gases; Vacuum cleaning; Window cleaning equipment and cradles; Compressed air; Treated water; Refrigerated stores.

Notes
(1)   The cost of each installation shall, where possible, be shown separately along with an indication of the specification.
(2)   Items deemed to be included under 'Refrigerated stores' comprise all plant required to provide refrigerated conditions (i.e. cooling towers, compressors, instrumentation and controls, cold room thermal insulation and vapour sealing, cold room doors, etc.) for cold rooms, refrigerated stores and the like other than that required for 'Space heating and air treatment' (5.F.8 and 5.F.9).

5.N Builder's work in connection with services

Builder's work in connection with mechanical and electrical services.

Notes
(1) The cost of builder's work in connection with each of the services elements (5.A to 5.M) shall, where possible, be shown separately.
(2) Where tank rooms, housings, and the like are included in the gross floor area, their component parts shall be analysed in detail under the appropriate elements. Where this is not the case the cost of such items shall be included here.

5.O Builder's profit and attendance on services

Builder's profit and attendance in connection with mechanical and electrical services.

Note
The cost of profit and attendance in connection with each of the services elements (5.A to 5.M) shall, where possible, be shown separately.

6.A Site works

6.A.1 Site preparation

Clearance and demolitions.
Preparatory earth works to form new contours.

6.A.2 Surface treatment

The cost of the following items shall be stated separately if possible:
Roads and associated footways.
Vehicle parks.
Paths and paved areas.
Playing fields.
Playgrounds.
Games courts.
Retaining walls.
Land drainage.
Landscape work.

6.A.3 Site enclosure and division

Gates and entrance.
Fencing, walling, and hedges.

6.A.4   Fittings and furniture

Notice boards, flag poles, seats, signs.

6.B   Drainage

Surface water drainage.
Foul drainage.
Sewage treatment.

Note
To include all drainage works (other than land drainage included with 'Surface treatment' (6.A.2)) outside the building to and including disposal point, connection to sewer or to treatment plant.

6.C   External services

6.C.1   Water mains

Main from existing supply up to external face of building.

6.C.2   Fire mains

Main from existing supply up to external face of building; fire hydrants.

6.C.3   Heating mains

Main from existing supply or heat source up to external face of building.

6.C.4   Gas mains

Main from existing supply up to external face of building.

6.C.5   Electric mains

Main from existing supply up to external face of building.

6.C.6   Site lighting

Distribution, fittings, and equipment.

6.C.7   Other mains and services

Mains relating to other service installations (each shown separately).

6.C.8   Builder's work in connection with external services

Builder's work in connection with external mechanical and electrical services: e.g. pits, trenches, ducts, etc.

Note
The cost of builder's work shall be stated separately for each of the sub-sections (6.C.1) to (6.C.7).

6.C.9   Builder's profit and attendance on external mechanical and electrical services

Note
The cost of profit and attendances shall be stated separately for each of the sub-sections (6.C.1) to (6.C.7).

6.D   Minor building work

6.D.1   Ancillary buildings

Separate minor buildings such as sub-stations, bicycle stores, horticultural buildings, and the like, inclusive of local engineering services.

6.D.2   Alterations to existing buildings

Alterations and minor additions, shoring, repair, and maintenance to existing buildings.

Preliminaries

Priced items in Preliminaries Bill and Summary but excluding contractors' price adjustments. Individual costs of the main preliminary items should be given.

Notes
(1)   Professional fees will not form part of the cost analysis.
(2)   Lump sum adjustments shall be spread pro rata amongst all elements of the building and external works based on all work excluding Prime Cost and Provisional Sums.

Appendix B
Discounting and Interest Formulae and Tables

The interest and discount formulae and tables in this appendix may be used in respect of weekly, monthly or yearly periods.

0.096% per week = 0.417% per month = 5% per annum
0.192% per week = 0.833% per month = 10% per annum
0.289% per week = 1.25% per month = 15% per annum
0.385% per week = 1.667% per month = 20% per annum

In the following formulae n represents the number of periods and i the interest rate expressed as a decimal fraction of the principal, e.g. 5% = 0.05.

Formula 1: Compound interest $(1+i)^n$

Formula 2: Future value of £1 invested at regular intervals $\dfrac{(1+i)^n - 1}{i}$

Formula 3: Present value of £1 $\dfrac{1}{(1+i)^n}$

Formula 4: Present value of £1 payable at regular intervals ('years purchase')

$$\frac{(1+i)^n - 1}{i(1+i)^n}$$

Formula 5: Annuity purchased by £1 $\dfrac{i(1+i)^n}{(1+i)^n - 1}$

Formula 6: Sinking fund $\dfrac{i}{(1+i)^n - 1}$

Note: No provision is made in Table 3 and in those that follow for repayment of the orginal sum by means of a sinking fund. When this is required the relevant value from Table 6 should be added.

Formulae 3, 5 and 6 are the reciprocals of formulae 1, 4 and 2 respectively.

Table 1 Compound interest.
Value at end of each period of £1 invested at beginning of period 1 and accumulating at compound interest from 1% to 30% per period.

Period	1% £	1.5% £	2% £	2.5% £	3% £	4% £	5% £	6% £	7% £	8% £
1	1.01	1.02	1.02	1.03	1.03	1.04	1.05	1.06	1.07	1.08
2	1.02	1.03	1.04	1.05	1.06	1.08	1.10	1.12	1.14	1.17
3	1.03	1.05	1.06	1.08	1.09	1.12	1.16	1.19	1.23	1.26
4	1.04	1.06	1.08	1.10	1.13	1.17	1.22	1.26	1.31	1.36
5	1.05	1.08	1.10	1.13	1.16	1.22	1.28	1.34	1.40	1.47
6	1.06	1.09	1.13	1.16	1.19	1.27	1.34	1.42	1.50	1.59
7	1.07	1.11	1.15	1.19	1.23	1.32	1.41	1.50	1.61	1.71
8	1.08	1.13	1.17	1.22	1.27	1.37	1.48	1.59	1.72	1.85
9	1.09	1.14	1.20	1.25	1.30	1.42	1.55	1.69	1.84	2.00
10	1.10	1.16	1.22	1.28	1.34	1.48	1.63	1.79	1.97	2.16
11	1.12	1.18	1.24	1.31	1.38	1.54	1.71	1.90	2.10	2.33
12	1.13	1.20	1.27	1.34	1.43	1.60	1.80	2.01	2.25	2.52
13	1.14	1.21	1.29	1.38	1.47	1.67	1.89	2.13	2.41	2.72
14	1.15	1.23	1.32	1.41	1.51	1.73	1.98	2.26	2.58	2.94
15	1.16	1.25	1.35	1.45	1.56	1.80	2.08	2.40	2.76	3.17
16	1.17	1.27	1.37	1.48	1.60	1.87	2.18	2.54	2.95	3.43
17	1.18	1.29	1.40	1.52	1.65	1.95	2.29	2.69	3.16	3.70
18	1.20	1.31	1.43	1.56	1.70	2.03	2.41	2.85	3.38	4.00
19	1.21	1.33	1.46	1.60	1.75	2.11	2.53	3.03	3.62	4.32
20	1.22	1.35	1.49	1.64	1.81	2.19	2.65	3.21	3.87	4.66

Cont.

Period	1% £	1.5% £	2% £	2.5% £	3% £	4% £	5% £	6% £	7% £	8% £
21	1.23	1.37	1.52	1.68	1.86	2.28	2.79	3.40	4.14	5.03
22	1.24	1.39	1.55	1.72	1.92	2.37	2.93	3.60	4.43	5.44
23	1.26	1.41	1.58	1.76	1.97	2.46	3.07	3.82	4.74	5.87
24	1.27	1.43	1.61	1.81	2.03	2.56	3.23	4.05	5.07	6.34
25	1.28	1.45	1.64	1.85	2.09	2.67	3.39	4.29	5.43	6.85
26	1.30	1.47	1.67	1.90	2.16	2.77	3.56	4.55	5.81	7.40
27	1.31	1.49	1.71	1.95	2.22	2.88	3.73	4.82	6.21	7.99
28	1.32	1.52	1.74	2.00	2.29	3.00	3.92	5.11	6.65	8.63
29	1.33	1.54	1.78	2.05	2.36	3.12	4.12	5.42	7.11	9.32
30	1.35	1.56	1.81	2.10	2.43	3.24	4.32	5.74	7.61	10.06
31	1.36	1.59	1.85	2.15	2.50	3.37	4.54	6.09	8.15	10.87
32	1.37	1.61	1.88	2.20	2.58	3.51	4.76	6.45	8.72	11.74
33	1.39	1.63	1.92	2.26	2.65	3.65	5.00	6.84	9.33	12.68
34	1.40	1.66	1.96	2.32	2.73	3.79	5.25	7.25	9.98	13.69
35	1.42	1.68	2.00	2.37	2.81	3.95	5.52	7.69	10.68	14.79
36	1.43	1.71	2.04	2.43	2.90	4.10	5.79	8.15	11.42	15.97
37	1.45	1.73	2.08	2.49	2.99	4.27	6.08	8.64	12.22	17.25
38	1.46	1.76	2.12	2.56	3.07	4.44	6.39	9.15	13.08	18.63
39	1.47	1.79	2.16	2.62	3.17	4.62	6.70	9.70	13.99	20.12
40	1.49	1.81	2.21	2.69	3.26	4.80	7.04	10.29	14.97	21.72

Cont.

Table 1 Compound interest.
Value at end of each period of £1 invested at beginning of period 1 and accumulating at compound interest from 1% to 30% per period.

Period	9% £	10% £	11% £	12% £	13% £	14% £	15% £	20% £	25% £	30% £
1	1.09	1.10	1.11	1.12	1.13	1.14	1.15	1.20	1.25	1.30
2	1.19	1.21	1.23	1.25	1.28	1.30	1.32	1.44	1.56	1.69
3	1.30	1.33	1.37	1.40	1.44	1.48	1.52	1.73	1.95	2.20
4	1.41	1.46	1.52	1.57	1.63	1.69	1.75	2.07	2.44	2.86
5	1.54	1.61	1.69	1.76	1.84	1.93	2.01	2.49	3.05	3.71
6	1.68	1.77	1.87	1.97	2.08	2.19	2.31	2.99	3.81	4.83
7	1.83	1.95	2.08	2.21	2.35	2.50	2.66	3.58	4.77	6.27
8	1.99	2.14	2.30	2.48	2.66	2.85	3.06	4.30	5.96	8.16
9	2.17	2.36	2.56	2.77	3.00	3.25	3.52	5.16	7.45	10.60
10	2.37	2.59	2.84	3.11	3.39	3.71	4.05	6.19	9.31	13.79
11	2.58	2.85	3.15	3.48	3.84	4.23	4.65	7.43	11.64	17.92
12	2.81	3.14	3.50	3.90	4.33	4.82	5.35	8.92	14.55	23.30
13	3.07	3.45	3.88	4.36	4.90	5.49	6.15	10.70	18.19	30.29
14	3.34	3.80	4.31	4.89	5.53	6.26	7.08	12.84	22.74	39.37
15	3.64	4.18	4.78	5.47	6.25	7.14	8.14	15.41	28.42	51.19
16	3.97	4.59	5.31	6.13	7.07	8.14	9.36	18.49	35.53	66.54
17	4.33	5.05	5.90	6.87	7.99	9.28	10.76	22.19	44.41	86.50
18	4.72	5.56	6.54	7.69	9.02	10.58	12.38	26.62	55.51	112.46
19	5.14	6.12	7.26	8.61	10.20	12.06	14.23	31.95	69.39	146.19
20	5.60	6.73	8.06	9.65	11.52	13.74	16.37	38.34	86.74	190.05

Cont.

Period	9% £	10% £	11% £	12% £	13% £	14% £	15% £	20% £	25% £	30% £
21	6.11	7.40	8.95	10.80	13.02	15.67	18.82	46.01	108.42	247.06
22	6.66	8.14	9.93	12.10	14.71	17.86	21.64	55.21	135.53	321.18
23	7.26	8.95	11.03	13.55	16.63	20.36	24.89	66.25	169.41	417.54
24	7.91	9.85	12.24	15.18	18.79	23.21	28.63	79.50	211.76	542.80
25	8.62	10.83	13.59	17.00	21.23	26.46	32.92	95.40	264.70	705.64
26	9.40	11.92	15.08	19.04	23.99	30.17	37.86	114.48	330.87	917.33
27	10.25	13.11	16.74	21.32	27.11	34.39	43.54	137.37	413.59	1192.53
28	11.17	14.42	18.58	23.88	30.63	39.20	50.07	164.84	516.99	1550.29
29	12.17	15.86	20.62	26.75	34.62	44.69	57.58	197.81	646.23	2015.38
30	13.27	17.45	22.89	29.86	39.12	50.95	66.21	237.38	807.79	2620.00
31	14.46	19.19	25.41	33.56	44.20	58.08	76.14	284.85	1009.74	3405.99
32	15.76	21.11	28.21	37.58	49.95	66.21	87.57	341.82	1262.18	4427.79
33	17.18	23.23	31.31	42.09	56.44	75.48	100.70	410.19	1577.72	5756.13
34	18.73	25.55	34.75	47.14	63.78	86.05	115.80	492.22	1972.15	7482.97
35	20.41	28.10	38.57	52.80	72.07	98.10	133.18	590.67	2465.19	9727.86
36	22.25	30.91	42.82	59.14	81.44	111.83	153.15	708.80	3081.49	12646.22
37	24.25	34.00	47.53	66.23	92.02	127.49	176.12	850.56	3851.86	16440.08
38	26.44	37.40	52.76	74.18	103.99	145.34	202.54	1020.67	4814.82	21372.11
39	28.82	41.14	58.56	83.08	117.51	165.69	232.92	1224.81	6018.53	27783.74
40	31.41	45.26	65.00	93.05	132.78	188.88	267.86	1469.77	7523.16	36118.86

Table 2 Future value of £1 invested at regular intervals.
Value of £1 invested regularly at end of each period (i.e. weekly, monthly, or yearly) accumulating at compound interest from 1% to 30% per period.

Period	1% £	1.5% £	2% £	2.5% £	3% £	4% £	5% £	6% £	7% £	8% £
1	1.00	1.00	1.00	1.00	1.00	1.00	1.00	1.00	1.00	1.00
2	2.01	2.02	2.02	2.03	2.03	2.04	2.05	2.06	2.07	2.08
3	3.03	3.05	3.06	3.08	3.09	3.12	3.15	3.18	3.21	3.25
4	4.06	4.09	4.12	4.15	4.18	4.25	4.31	4.37	4.44	4.51
5	5.10	5.15	5.20	5.26	5.31	5.42	5.53	5.64	5.75	5.87
6	6.15	6.23	6.31	6.39	6.47	6.63	6.80	6.98	7.15	7.34
7	7.21	7.32	7.43	7.55	7.66	7.90	8.14	8.39	8.65	8.92
8	8.29	8.43	8.58	8.74	8.89	9.21	9.55	9.90	10.26	10.64
9	9.37	9.56	9.75	9.95	10.16	10.58	11.03	11.49	11.98	12.49
10	10.46	10.70	10.95	11.20	11.46	12.01	12.58	13.18	13.82	14.49
11	11.57	11.86	12.17	12.48	12.81	13.49	14.21	14.97	15.78	16.65
12	12.68	13.04	13.41	13.80	14.19	15.03	15.92	16.87	17.89	18.98
13	13.81	14.24	14.68	15.14	15.62	16.63	17.71	18.88	20.14	21.50
14	14.95	15.45	15.97	16.52	17.09	18.29	19.60	21.02	22.55	24.21
15	16.10	16.68	17.29	17.93	18.60	20.02	21.58	23.28	25.13	27.15
16	17.26	17.93	18.64	19.38	20.16	21.82	23.66	25.67	27.89	30.32
17	18.43	19.20	20.01	20.86	21.76	23.70	25.84	28.21	30.84	33.75
18	19.61	20.49	21.41	22.39	23.41	25.65	28.13	30.91	34.00	37.45
19	20.81	21.80	22.84	23.95	25.12	27.67	30.54	33.76	37.38	41.45
20	22.02	23.12	24.30	35.54	26.87	29.78	33.07	36.79	41.00	45.76

Cont.

Period	1% £	1.5% £	2% £	2.5% £	3% £	4% £	5% £	6% £	7% £	8% £
21	23.24	24.47	25.78	27.18	28.68	31.97	35.72	39.99	44.87	50.42
22	24.47	25.84	27.30	28.86	30.54	34.25	38.51	43.39	49.01	55.46
23	25.72	27.23	28.84	30.58	32.45	36.62	41.43	47.00	53.44	60.89
24	26.97	28.63	30.42	32.35	34.43	39.08	44.50	50.82	58.18	66.76
25	28.24	30.06	32.03	34.16	36.46	41.65	47.73	54.86	63.25	73.11
26	29.53	31.51	33.67	36.01	38.55	44.31	51.11	59.16	68.68	79.95
27	30.82	32.99	35.34	37.91	40.71	47.08	54.67	63.71	74.48	87.35
28	32.13	34.48	37.05	39.86	42.93	49.97	58.40	68.53	80.70	95.34
29	33.45	36.00	38.79	41.86	45.22	52.97	62.32	73.64	87.35	103.97
30	34.78	37.54	40.57	43.90	47.58	56.08	66.44	79.06	94.46	113.28
31	36.13	39.10	42.38	46.00	50.00	59.33	70.76	84.80	102.07	123.35
32	37.49	40.69	44.23	48.15	52.50	62.70	75.30	90.89	110.22	134.21
33	38.87	42.30	46.11	50.35	55.08	66.21	80.06	97.34	118.93	145.95
34	40.26	43.93	48.03	52.61	57.73	69.86	85.07	104.18	128.26	158.63
35	41.66	45.59	49.99	54.93	60.46	73.65	90.32	111.43	138.24	172.32
36	43.08	47.28	51.99	57.30	63.28	77.60	95.84	119.12	148.91	187.10
37	44.51	48.99	54.03	59.73	66.17	81.70	101.63	127.27	160.34	203.07
38	45.95	50.72	56.11	62.23	69.16	85.97	107.71	135.90	172.56	220.32
39	47.41	52.48	58.24	64.78	72.23	90.41	114.10	145.06	185.64	238.94
40	48.89	54.27	60.40	67.40	75.40	95.03	120.80	154.76	199.64	259.06

Cont.

Table 2 Future value of £1 invested at regular intervals.
Value of £1 invested regularly at end of each period (i.e. weekly, monthly, or yearly) accumulating at compound interest from 1% to 30% per period.

Period	9% £	10% £	11% £	12% £	13% £	14% £	15% £	20% £	25% £	30% £
1	1.00	1.00	1.00	1.00	1.00	1.00	1.00	1.00	1.00	1.00
2	2.09	2.10	2.11	2.12	2.13	2.14	2.15	2.20	2.25	2.30
3	3.28	3.31	3.34	3.37	3.41	3.44	3.47	3.64	3.81	3.99
4	4.57	4.64	4.71	4.78	4.85	4.92	4.99	5.37	5.77	6.19
5	5.98	6.11	6.23	6.35	6.48	6.61	6.74	7.44	8.21	9.04
6	7.52	7.72	7.91	8.12	8.32	8.54	8.75	9.93	11.26	12.76
7	9.20	9.49	9.78	10.09	10.40	10.73	11.07	12.92	15.07	17.58
8	11.03	11.44	11.86	12.30	12.76	13.23	13.73	16.50	19.84	23.86
9	13.02	13.58	14.16	14.78	15.42	16.09	16.79	20.80	25.80	32.01
10	15.19	15.94	16.72	17.55	18.42	19.34	20.30	25.96	33.25	42.62
11	17.56	18.53	19.56	20.65	21.81	23.04	24.35	32.15	42.57	56.41
12	20.14	21.38	22.71	24.13	25.65	27.27	29.00	39.58	54.21	74.33
13	22.95	24.52	26.21	28.03	29.98	32.09	34.35	48.50	68.76	97.63
14	26.02	27.97	30.09	32.39	34.88	37.58	40.50	59.20	86.95	127.91
15	29.36	31.77	34.41	37.28	40.42	43.84	47.58	72.04	109.69	167.29
16	33.00	35.95	39.19	42.75	46.67	50.98	55.72	87.44	138.11	218.47
17	36.97	40.54	44.50	48.88	53.74	59.12	65.08	105.93	173.64	285.01
18	41.30	45.60	50.40	55.75	61.73	68.39	75.84	128.12	218.04	371.52
19	46.02	51.16	56.94	63.44	70.75	78.97	88.21	154.74	273.56	483.97
20	51.16	57.27	64.20	72.05	80.95	91.02	102.44	186.69	342.94	630.17

Cont.

Period	9% £	10% £	11% £	12% £	13% £	14% £	15% £	20% £	25% £	30% £
21	56.76	64.00	72.27	81.70	92.47	104.72	118.81	225.03	429.68	820.22
22	62.87	71.40	81.21	92.50	105.49	120.44	137.63	271.03	538.10	1067.28
23	69.53	79.54	91.15	104.60	120.20	138.30	159.28	326.24	673.63	1388.46
24	76.79	88.50	102.17	118.16	136.83	158.66	184.17	392.48	843.03	1806.00
25	84.70	98.35	114.41	133.33	155.62	181.87	212.79	471.98	1054.79	2348.80
26	93.32	109.18	128.00	150.33	176.85	208.33	245.71	567.38	1319.49	3054.44
27	102.72	121.10	143.08	169.37	200.84	238.50	283.57	681.85	1650.36	3971.78
28	112.97	134.21	159.82	190.70	227.95	272.89	327.10	819.22	2063.95	5164.31
29	124.14	148.63	178.40	214.58	258.58	312.09	377.17	984.07	2580.94	6714.60
30	136.31	164.49	199.02	241.33	293.20	356.79	434.75	1181.88	3227.17	8729.99
31	149.58	181.94	221.91	271.29	332.32	407.74	500.96	1419.26	4034.97	11349.98
32	164.04	201.14	247.32	304.85	376.52	465.82	577.10	1704.11	5044.71	14755.98
33	179.80	222.53	275.53	342.43	426.46	532.04	664.67	2045.93	6306.89	19183.77
34	196.98	245.48	306.84	384.52	482.90	607.52	765.37	2456.12	7884.61	24939.90
35	215.71	271.02	341.59	431.66	546.68	693.57	881.17	2948.34	9856.76	32422.87
36	236.12	299.13	380.16	484.46	618.75	791.67	1014.35	3539.01	12321.95	42150.73
37	258.38	330.04	422.98	543.60	700.19	903.51	1167.50	4247.81	15403.44	54796.95
38	282.63	364.04	470.51	609.83	792.21	1031.00	1343.62	5098.37	19255.30	71237.03
39	309.07	401.45	523.27	684.01	896.20	1176.34	1546.17	6119.05	24070.12	92609.14
40	337.88	442.59	581.83	767.09	1013.70	1342.03	1779.09	7343.86	30088.66	120392.88

Table 3   Present value of £1.

Present value of £1 payable (or receivable) at end of any period 1 to 40, discounted at interest rates from 1% to 30% per period. Values shown in pence.

Period	1% p	1.5% p	2% p	2.5% p	3% p	4% p	5% p	6% p	7% p	8% p
1	99.0	98.5	98.0	97.6	97.1	96.2	95.2	94.3	93.5	92.6
2	98.0	97.1	96.1	95.2	94.3	92.5	90.7	89.0	87.3	85.7
3	97.1	95.6	94.2	92.9	91.5	88.9	86.4	84.0	81.6	79.4
4	96.1	94.2	92.4	90.6	88.8	85.5	82.3	79.2	76.3	73.5
5	95.1	92.8	90.6	88.4	86.3	82.2	78.4	74.7	71.3	68.1
6	94.2	91.5	88.8	86.2	83.7	79.0	74.6	70.5	66.6	63.0
7	93.3	90.1	87.1	84.1	81.3	76.0	71.1	66.5	62.3	58.3
8	92.3	88.8	85.3	82.1	78.9	73.1	67.7	62.7	58.2	54.0
9	91.4	87.5	83.7	80.1	76.6	70.3	64.5	59.2	54.4	50.0
10	90.5	86.2	82.0	78.1	74.4	67.6	61.4	55.8	50.8	46.3
11	89.6	84.9	80.4	76.2	72.2	65.0	58.5	52.7	47.5	42.9
12	88.7	83.6	78.8	74.4	70.1	62.5	55.7	49.7	44.4	39.7
13	87.9	82.4	77.3	72.5	68.1	60.1	53.0	46.9	41.5	36.8
14	87.0	81.2	75.8	70.8	66.1	57.7	50.5	44.2	38.8	34.0
15	86.1	80.0	74.3	69.0	64.2	55.5	48.1	41.7	36.2	31.5
16	85.3	78.8	72.8	67.4	62.3	53.4	45.8	39.4	33.9	29.2
17	84.4	77.6	71.4	65.7	60.5	51.3	43.6	37.1	31.7	27.0
18	83.6	76.5	70.0	64.1	58.7	49.4	41.6	35.0	29.6	25.0
19	82.8	75.4	68.6	62.6	57.0	47.5	39.6	33.1	27.7	23.2
20	82.0	74.2	67.3	61.0	55.4	45.6	37.7	31.2	25.8	21.5

Cont.

Period	1%	1.5%	2%	2.5%	3%	4%	5%	6%	7%	8%
	p	p	p	p	p	p	p	p	p	p
21	81.1	73.1	66.0	59.5	53.8	43.9	35.9	29.4	24.2	19.9
22	80.3	72.1	64.7	58.1	52.2	42.2	34.2	27.8	22.6	18.4
23	79.5	71.0	63.4	56.7	50.7	40.6	32.6	26.2	21.1	17.0
24	78.8	70.0	62.2	55.3	49.2	39.0	31.0	24.7	19.7	15.8
25	78.0	68.9	61.0	53.9	47.8	37.5	29.5	23.3	18.4	14.6
26	77.2	67.9	59.8	52.6	46.4	36.1	28.1	22.0	17.2	13.5
27	76.4	66.9	58.6	51.3	45.0	34.7	26.8	20.7	16.1	12.5
28	75.7	65.9	57.4	50.1	43.7	33.3	25.5	19.6	15.0	11.6
29	74.9	64.9	56.3	48.9	42.4	32.1	24.3	18.5	14.1	10.7
30	74.2	64.0	55.2	47.7	41.2	30.8	23.1	17.4	13.1	9.9
31	73.5	63.0	54.1	46.5	40.0	29.6	22.0	16.4	12.3	9.2
32	72.7	62.1	53.1	45.4	38.8	28.5	21.0	15.5	11.5	8.5
33	72.0	61.2	52.0	44.3	37.7	27.4	20.0	14.6	10.7	7.9
34	71.3	60.3	51.0	43.2	36.6	26.4	19.0	13.8	10.0	7.3
35	70.6	59.4	50.0	42.1	35.5	25.3	18.1	13.0	9.4	6.8
36	69.9	58.5	49.0	41.1	34.5	24.4	17.3	12.3	8.8	6.3
37	69.2	57.6	48.1	40.1	33.5	23.4	16.4	11.6	8.2	5.8
38	68.5	56.8	47.1	39.1	32.5	22.5	15.7	10.9	7.6	5.4
39	67.8	56.0	46.2	38.2	31.6	21.7	14.9	10.3	7.1	5.0
40	67.2	55.1	45.3	37.2	30.7	20.8	14.2	9.7	6.7	4.6

Cont.

Table 3 Present value of £1.
Present value of £1 payable (or receivable) at end of any period 1 to 40, discounted at interest rates from 1% to 30% per period. Values shown in pence.

Period	9% p	10% p	11% p	12% p	13% p	14% p	15% p	20% p	25% p	30% p
1	91.7	90.9	90.1	89.3	88.5	87.7	87.0	83.3	80.0	76.9
2	84.2	82.6	81.2	79.7	78.3	76.9	75.6	69.4	64.0	59.2
3	77.2	75.1	73.1	71.2	69.3	67.5	65.8	57.9	51.2	45.5
4	70.8	68.3	65.9	63.6	61.3	59.2	57.2	48.2	41.0	35.0
5	65.0	62.1	59.3	56.7	54.3	51.9	49.7	40.2	32.8	26.9
6	59.6	56.4	53.5	50.7	48.0	45.6	43.2	33.5	26.2	20.7
7	54.7	51.3	48.2	45.2	42.5	40.0	37.6	27.9	21.0	15.9
8	50.2	46.7	43.4	40.4	37.6	35.1	32.7	23.3	16.8	12.3
9	46.0	42.4	39.1	36.1	33.3	30.8	28.4	19.4	13.4	9.4
10	42.2	38.6	35.2	32.2	29.5	27.0	24.7	16.2	10.7	7.3
11	38.8	35.0	31.7	28.7	26.1	23.7	21.5	13.5	8.6	5.6
12	35.6	31.9	28.6	25.7	23.1	20.8	18.7	11.2	6.9	4.3
13	32.6	29.0	25.8	22.9	20.4	18.2	16.3	9.3	5.5	3.3
14	29.9	26.3	23.2	20.5	18.1	16.0	14.1	7.8	4.4	2.5
15	27.5	23.9	20.9	18.3	16.0	14.0	12.3	6.5	3.5	2.0
16	25.2	21.8	18.8	16.3	14.1	12.3	10.7	5.4	2.8	1.5
17	23.1	19.8	17.0	14.6	12.5	10.8	9.3	4.5	2.3	1.2
18	21.2	18.0	15.3	13.0	11.1	9.5	8.1	3.8	1.8	0.9
19	19.4	16.4	13.8	11.6	9.8	8.3	7.0	3.1	1.4	0.7
20	17.8	14.9	12.4	10.4	8.7	7.3	6.1	2.6	1.2	0.5

Cont.

Period	9%	10%	11%	12%	13%	14%	15%	20%	25%	30%
	p	p	p	p	p	p	p	p	p	p
21	16.4	13.5	11.2	9.3	7.7	6.4	5.3	2.2	0.9	0.4
22	15.0	12.3	10.1	8.3	6.8	5.6	4.6	1.8	0.7	0.3
23	13.8	11.2	9.1	7.4	6.0	4.9	4.0	1.5	0.6	0.2
24	12.6	10.2	8.2	6.6	5.3	4.3	3.5	1.3	0.5	0.2
25	11.6	9.2	7.4	5.9	4.7	3.8	3.0	1.0	0.4	0.1
26	10.6	8.4	6.6	5.3	4.2	3.3	2.6	0.9	0.3	0.1
27	9.8	7.6	6.0	4.7	3.7	2.9	2.3	0.7	0.2	0.1
28	9.0	6.9	5.4	4.2	3.3	2.6	2.0	0.6	0.2	0.1
29	8.2	6.3	4.8	3.7	2.9	2.2	1.7	0.5	0.2	
30	7.5	5.7	4.4	3.3	2.6	2.0	1.5	0.4	0.2	
31	6.9	5.2	3.9	3.0	2.3	1.7	1.3	0.4	0.1	
32	6.3	4.7	3.5	2.7	2.0	1.5	1.1	0.3	0.1	
33	5.8	4.3	3.2	2.4	1.8	1.3	1.0	0.2	0.1	
34	5.3	3.9	2.9	2.1	1.6	1.2	0.9	0.2	0.1	
35	4.9	3.6	2.6	1.9	1.4	1.0	0.8	0.2	0.1	
36	4.5	3.2	2.3	1.7	1.2	0.9	0.7	0.2		
37	4.1	2.9	2.1	1.5	1.1	0.8	0.6	0.1		
38	3.8	2.7	1.9	1.3	1.0	0.7	0.6	0.1		
39	3.5	2.4	1.7	1.2	0.9	0.6	0.5	0.1		
40	3.2	2.2	1.5	1.1	0.8	0.5	0.4	0.1		

Table 4 Present value of £1 payable at regular intervals ('years purchase').
Present value of £1 payable (or receivable) regularly at end of each period (i.e. weekly, monthly, or yearly) discounted at interest rates from 1% to 30% per period.

Period	1% £	1.5% £	2% £	2.5% £	3% £	4% £	5% £	6% £	7% £	8% £
1	0.99	0.99	0.98	0.98	0.97	0.96	0.95	0.94	0.93	0.93
2	1.97	1.96	1.94	1.93	1.91	1.89	1.86	1.83	1.81	1.78
3	2.94	2.91	2.88	2.86	2.83	2.78	2.72	2.67	2.62	2.58
4	3.90	3.85	3.81	3.76	3.72	3.63	3.55	3.47	3.39	3.31
5	4.85	4.78	4.71	4.65	4.58	4.45	4.33	4.21	4.10	3.99
6	5.80	5.70	5.60	5.51	5.42	5.24	5.08	4.92	4.77	4.62
7	6.73	6.60	6.47	6.35	6.23	6.00	5.79	5.58	5.39	5.21
8	7.65	7.49	7.33	7.17	7.02	6.73	6.46	6.21	5.97	5.75
9	8.57	8.36	8.16	7.97	7.79	7.44	7.11	6.80	6.52	6.25
10	9.47	9.22	8.98	8.75	8.53	8.11	7.72	7.36	7.02	6.71
11	10.37	10.07	9.79	9.51	9.25	8.76	8.31	7.89	7.50	7.14
12	11.26	10.91	10.58	10.26	9.95	9.39	8.86	8.38	7.94	7.54
13	12.13	11.73	11.35	10.98	10.63	9.99	9.39	8.85	8.36	7.90
14	13.00	12.54	12.11	11.69	11.30	10.56	9.90	9.29	8.75	8.24
15	13.87	13.34	12.85	12.38	11.94	11.12	10.38	9.71	9.11	8.56
16	14.72	14.13	13.58	13.06	12.56	11.65	10.84	10.11	9.45	8.85
17	15.56	14.91	14.29	13.71	13.17	12.17	11.27	10.48	9.76	9.12
18	16.40	15.67	14.99	14.35	13.75	12.66	11.69	10.83	10.06	9.37
19	17.23	16.43	15.68	14.98	14.32	13.13	12.09	11.16	10.34	9.60
20	18.05	17.17	16.35	15.59	14.88	13.59	12.46	11.47	10.59	9.82

Cont.

Period	1% £	1.5% £	2% £	2.5% £	3% £	4% £	5% £	6% £	7% £	8% £
21	18.86	17.90	17.01	16.18	15.42	14.03	12.82	11.76	10.84	10.02
22	19.66	18.62	17.66	16.77	15.94	14.45	13.16	12.04	11.06	10.20
23	20.46	19.33	18.29	17.33	16.44	14.86	13.49	12.30	11.27	10.37
24	21.24	20.03	18.91	17.88	16.94	15.25	13.80	12.55	11.47	10.53
25	22.02	20.72	19.52	18.42	17.41	15.62	14.09	12.78	11.65	10.67
26	22.80	21.40	20.12	18.95	17.88	15.98	14.38	13.00	11.83	10.81
27	23.56	22.07	20.71	19.46	18.33	16.33	14.64	13.21	11.99	10.94
28	24.32	22.73	21.28	19.96	18.76	16.66	14.90	13.41	12.14	11.05
29	25.07	23.38	21.84	20.45	19.19	16.98	15.14	13.59	12.28	11.16
30	25.81	24.02	22.40	20.93	19.60	17.29	15.37	13.76	12.41	11.26
31	26.54	24.65	22.94	21.40	20.00	17.59	15.59	13.93	12.53	11.35
32	27.27	25.27	23.47	21.85	20.39	17.87	15.80	14.08	12.65	11.43
33	27.99	25.88	23.99	22.29	20.77	18.15	16.00	14.23	12.75	11.51
34	28.70	26.48	24.50	22.72	21.13	18.41	16.19	14.37	12.85	11.59
35	29.41	27.08	25.00	23.15	21.49	18.66	16.37	14.50	12.95	11.65
36	30.11	27.66	25.49	23.56	21.83	18.91	16.55	14.62	13.04	11.72
37	30.80	28.24	25.97	23.96	22.17	19.14	16.71	14.74	13.12	11.78
38	31.48	28.81	26.44	24.35	22.49	19.37	16.87	14.85	13.19	11.83
39	32.16	29.36	26.90	24.73	22.81	19.58	17.02	14.95	13.26	11.88
40	32.83	29.92	27.36	25.10	23.11	19.79	17.16	15.05	13.33	11.92

Cont.

Table 4 Present value of £1 payable at regular intervals ('years purchase').
Present value of £1 payable (or receivable) regularly at end of each period (i.e. weekly, monthly, or yearly) discounted at interest rates from 1% to 30% per period.

Period	9% £	10% £	11% £	12% £	13% £	14% £	15% £	20% £	25% £	30% £
1	0.92	0.91	0.90	0.89	0.88	0.88	0.87	0.83	0.80	0.77
2	1.76	1.74	1.71	1.69	1.67	1.65	1.63	1.53	1.44	1.36
3	2.53	2.49	2.44	2.40	2.36	2.32	2.28	2.11	1.95	1.82
4	3.24	3.17	3.10	3.04	2.97	2.91	2.85	2.59	2.36	2.17
5	3.89	3.79	3.70	3.60	3.52	3.43	3.35	2.99	2.69	2.44
6	4.49	4.36	4.23	4.11	4.00	3.89	3.78	3.33	2.95	2.64
7	5.03	4.87	4.71	4.56	4.42	4.29	4.16	3.60	3.16	2.80
8	5.53	5.33	5.15	4.97	4.80	4.64	4.49	3.84	3.33	2.92
9	6.00	5.76	5.54	5.33	5.13	4.95	4.77	4.03	3.46	3.02
10	6.42	6.14	5.89	5.65	5.43	5.22	5.02	4.19	3.57	3.09
11	6.81	6.50	6.21	5.94	5.69	5.45	5.23	4.33	3.66	3.15
12	7.16	6.81	6.49	6.19	5.92	5.66	5.42	4.44	3.73	3.19
13	7.49	7.10	6.75	6.42	6.12	5.84	5.58	4.53	3.78	3.22
14	7.79	7.37	6.98	6.63	6.30	6.00	5.72	4.61	3.82	3.25
15	8.06	7.61	7.19	6.81	6.46	6.14	5.85	4.68	3.86	3.27
16	8.31	7.82	7.38	6.97	6.60	6.27	5.95	4.73	3.89	3.28
17	8.54	8.02	7.55	7.12	6.73	6.37	6.05	4.77	3.91	3.29
18	8.76	8.20	7.70	7.25	6.84	6.47	6.13	4.81	3.93	3.30
19	8.95	8.36	7.84	7.37	6.94	6.55	6.20	4.84	3.94	3.31
20	9.13	8.51	7.96	7.47	7.02	6.62	6.26	4.87	3.95	3.32

Cont.

Period	9% £	10% £	11% £	12% £	13% £	14% £	15% £	20% £	25% £	30% £
21	9.29	8.65	8.08	7.56	7.10	6.69	6.31	4.89	3.96	3.32
22	9.44	8.77	8.18	7.64	7.17	6.74	6.36	4.91	3.97	3.32
23	9.58	8.88	8.27	7.72	7.23	6.79	6.40	4.92	3.98	3.33
24	9.71	8.98	8.35	7.78	7.28	6.84	6.43	4.94	3.98	3.33
25	9.82	9.08	8.42	7.84	7.33	6.87	6.46	4.95	3.98	3.33
26	9.93	9.16	8.49	7.90	7.37	6.91	6.49	4.96	3.99	3.33
27	10.03	9.24	8.55	7.94	7.41	6.94	6.51	4.96	3.99	3.33
28	10.12	9.31	8.60	7.98	7.44	6.96	6.53	4.97	3.99	3.33
29	10.20	9.37	8.65	8.02	7.47	6.98	6.55	4.97	3.99	3.33
30	10.27	9.43	8.69	8.06	7.50	7.00	6.57	4.98	4.00	3.33
31	10.34	9.48	8.73	8.08	7.52	7.02	6.58	4.98	4.00	3.33
32	10.41	9.53	8.77	8.11	7.54	7.03	6.59	4.99	4.00	3.33
33	10.46	9.57	8.80	8.14	7.56	7.05	6.60	4.99	4.00	
34	10.52	9.61	8.83	8.16	7.57	7.06	6.61	4.99	4.00	
35	10.57	9.64	8.86	8.18	7.59	7.07	6.62	4.99	4.00	
36	10.61	9.68	8.88	8.19	7.60	7.08	6.62	4.99	4.00	
37	10.65	9.71	8.90	8.21	7.61	7.09	6.63	4.99		
38	10.69	9.73	8.92	8.22	7.62	7.09	6.63	5.00		
39	10.73	9.76	8.94	8.23	7.63	7.10	6.64	5.00		
40	10.76	9.78	8.95	8.24	7.63	7.11	6.64	5.00		

Table 5 Annuity £1 will purchase (annual equivalent).
Annual equivalent of £1 invested at the beginning of the period or the annuity purchased by a lump sum payment of £1 at rates of interest from 1% to 30% per period.

Period	1% £	1.5% £	2% £	2.5% £	3% £	4% £	5% £	6% £	7% £	8% £
1	1.010	1.015	1.020	1.025	1.030	1.040	1.050	1.060	1.070	1.080
2	0.508	0.511	0.515	0.519	0.523	0.530	0.538	0.545	0.553	0.561
3	0.340	0.343	0.347	0.350	0.354	0.360	0.367	0.374	0.381	0.388
4	0.256	0.259	0.263	0.266	0.269	0.275	0.282	0.289	0.295	0.302
5	0.206	0.209	0.212	0.215	0.218	0.225	0.231	0.237	0.244	0.250
6	0.173	0.176	0.179	0.182	0.185	0.191	0.197	0.203	0.210	0.216
7	0.149	0.152	0.155	0.157	0.161	0.167	0.173	0.179	0.186	0.192
8	0.131	0.134	0.137	0.139	0.142	0.149	0.155	0.161	0.167	0.174
9	0.117	0.120	0.123	0.125	0.128	0.134	0.141	0.147	0.153	0.160
10	0.106	0.108	0.111	0.114	0.117	0.123	0.130	0.136	0.142	0.149
11	0.096	0.099	0.102	0.105	0.108	0.114	0.120	0.127	0.133	0.140
12	0.089	0.092	0.095	0.097	0.100	0.107	0.113	0.119	0.126	0.133
13	0.082	0.085	0.088	0.091	0.094	0.100	0.106	0.113	0.120	0.127
14	0.077	0.080	0.083	0.086	0.089	0.095	0.101	0.108	0.114	0.121
15	0.072	0.075	0.078	0.081	0.084	0.090	0.096	0.103	0.110	0.117
16	0.068	0.071	0.074	0.077	0.080	0.086	0.092	0.099	0.106	0.113
17	0.064	0.067	0.070	0.073	0.076	0.082	0.089	0.095	0.102	0.110
18	0.061	0.064	0.067	0.070	0.073	0.079	0.086	0.092	0.099	0.107
19	0.058	0.061	0.064	0.067	0.070	0.076	0.083	0.090	0.097	0.104
20	0.055	0.058	0.061	0.064	0.067	0.074	0.080	0.087	0.094	0.102

Cont.

Period	1% £	1.5% £	2% £	2.5% £	3% £	4% £	5% £	6% £	7% £	8% £
21	0.053	0.056	0.059	0.062	0.065	0.071	0.078	0.085	0.092	0.100
22	0.051	0.054	0.057	0.060	0.063	0.069	0.076	0.083	0.090	0.098
23	0.049	0.052	0.055	0.058	0.061	0.067	0.074	0.081	0.089	0.096
24	0.047	0.050	0.053	0.056	0.059	0.066	0.072	0.080	0.087	0.095
25	0.045	0.048	0.051	0.054	0.057	0.064	0.071	0.078	0.086	0.094
26	0.044	0.047	0.050	0.053	0.056	0.063	0.070	0.077	0.085	0.093
27	0.042	0.045	0.048	0.051	0.055	0.061	0.068	0.076	0.083	0.091
28	0.041	0.044	0.047	0.050	0.053	0.060	0.067	0.075	0.082	0.090
29	0.040	0.043	0.046	0.049	0.052	0.059	0.066	0.074	0.081	0.090
30	0.039	0.042	0.045	0.048	0.051	0.058	0.065	0.073	0.081	0.089
31	0.038	0.041	0.044	0.047	0.050	0.057	0.064	0.072	0.080	0.088
32	0.037	0.040	0.043	0.046	0.049	0.056	0.063	0.071	0.079	0.087
33	0.036	0.039	0.042	0.045	0.048	0.055	0.062	0.070	0.078	0.087
34	0.035	0.038	0.041	0.044	0.047	0.054	0.062	0.070	0.078	0.086
35	0.034	0.037	0.040	0.043	0.047	0.054	0.061	0.069	0.077	0.086
36	0.033	0.036	0.039	0.042	0.046	0.053	0.060	0.068	0.077	0.085
37	0.032	0.035	0.039	0.042	0.045	0.052	0.060	0.068	0.076	0.085
38	0.032	0.035	0.038	0.041	0.044	0.052	0.059	0.067	0.076	0.085
39	0.031	0.034	0.037	0.040	0.044	0.051	0.059	0.067	0.075	0.084
40	0.030	0.033	0.037	0.040	0.043	0.051	0.058	0.066	0.075	0.084

Cont.

Table 5   Annuity £1 will purchase (annual equivalent).
Annual equivalent of £1 invested at the beginning of the period or the annuity purchased by a lump sum payment of £1 at rates of interest from 1% to 30% per period.

Period	9% £	10% £	11% £	12% £	13% £	14% £	15% £	20% £	25% £	30% £
1	1.090	1.100	1.110	1.120	1.130	1.140	1.150	1.200	1.250	1.300
2	0.568	0.576	0.584	0.592	0.599	0.607	0.615	0.655	0.694	0.735
3	0.395	0.402	0.409	0.416	0.424	0.431	0.438	0.475	0.512	0.551
4	0.309	0.315	0.322	0.329	0.336	0.343	0.350	0.386	0.423	0.462
5	0.257	0.264	0.271	0.277	0.284	0.291	0.298	0.334	0.372	0.411
6	0.223	0.230	0.236	0.243	0.250	0.257	0.264	0.301	0.339	0.378
7	0.199	0.205	0.212	0.219	0.226	0.233	0.240	0.277	0.316	0.357
8	0.181	0.187	0.194	0.201	0.208	0.216	0.223	0.261	0.300	0.342
9	0.167	0.174	0.181	0.188	0.195	0.202	0.210	0.248	0.289	0.331
10	0.156	0.163	0.170	0.177	0.184	0.192	0.199	0.239	0.280	0.323
11	0.147	0.154	0.161	0.168	0.176	0.183	0.191	0.231	0.273	0.318
12	0.140	0.147	0.154	0.161	0.169	0.177	0.184	0.225	0.268	0.313
13	0.134	0.141	0.148	0.156	0.163	0.171	0.179	0.221	0.265	0.310
14	0.128	0.136	0.143	0.151	0.159	0.167	0.175	0.217	0.262	0.308
15	0.124	0.131	0.139	0.147	0.155	0.163	0.171	0.214	0.259	0.306
16	0.120	0.128	0.136	0.143	0.151	0.160	0.168	0.211	0.257	0.305
17	0.117	0.125	0.132	0.140	0.149	0.157	0.165	0.209	0.256	0.304
18	0.114	0.122	0.130	0.138	0.146	0.155	0.163	0.208	0.255	0.303
19	0.112	0.120	0.128	0.136	0.144	0.153	0.161	0.206	0.254	0.302
20	0.110	0.117	0.126	0.134	0.142	0.151	0.160	0.205	0.253	0.302

Cont.

Period	9% £	10% £	11% £	12% £	13% £	14% £	15% £	20% £	25% £	30% £
21	0.108	0.116	0.124	0.132	0.141	0.150	0.158	0.204	0.252	0.301
22	0.106	0.114	0.122	0.131	0.139	0.148	0.157	0.204	0.252	0.301
23	0.104	0.113	0.121	0.130	0.138	0.147	0.156	0.203	0.251	0.301
24	0.103	0.111	0.128	0.128	0.137	0.146	0.155	0.203	0.251	0.301
25	0.102	0.110	0.119	0.127	0.136	0.145	0.155	0.202	0.251	0.300
26	0.101	0.109	0.118	0.127	0.136	0.145	0.154	0.202	0.251	0.300
27	0.100	0.108	0.117	0.126	0.135	0.144	0.154	0.201	0.251	0.300
28	0.099	0.107	0.116	0.125	0.134	0.144	0.153	0.201	0.250	0.300
29	0.098	0.107	0.116	0.125	0.134	0.143	0.153	0.201	0.250	0.300
30	0.097	0.106	0.115	0.124	0.133	0.143	0.152	0.201	0.250	0.300
31	0.097	0.105	0.115	0.124	0.133	0.142	0.152	0.201	0.250	0.300
32	0.096	0.105	0.114	0.123	0.133	0.142	0.152	0.201	0.250	0.300
33	0.096	0.104	0.114	0.123	0.132	0.142	0.152	0.200	0.250	0.300
34	0.095	0.104	0.113	0.123	0.132	0.142	0.151	0.200	0.250	0.300
35	0.095	0.104	0.113	0.122	0.132	0.141	0.151	0.200	0.250	0.300
36	0.094	0.103	0.113	0.122	0.132	0.141	0.151	0.200	0.250	0.300
37	0.094	0.103	0.112	0.122	0.131	0.141	0.151	0.200	0.250	0.300
38	0.94	0.103	0.112	0.122	0.131	0.141	0.151	0.200	0.250	0.300
39	0.093	0.102	0.112	0.121	0.131	0.141	0.151	0.200	0.250	0.300
40	0.093	0.102	0.112	0.121	0.131	0.141	0.151	0.200	0.250	0.300

Table 6 Sinking fund.

Amount (in pence) which has to be invested regularly at the end of each period (i.e. weekly, monthly, or yearly) in order to accumulate to £1 by the end of the chosen term; at compound interest of from 0.096% to 12% per period.

Period	1% p	1.5% p	2% p	3% p	4% p	5% p	6% p	8% p	10% p	12% p
2	49.8	49.6	49.5	49.3	49.0	48.8	48.5	48.1	47.6	47.2
3	33.0	32.8	32.7	32.4	32.0	31.7	31.4	30.8	30.2	29.6
4	24.6	24.4	24.3	23.9	23.5	23.2	22.9	22.2	21.5	20.9
5	19.6	19.4	19.2	18.8	18.5	18.1	17.7	17.0	16.4	15.7
6	16.3	16.1	15.9	15.5	15.1	14.7	14.3	13.6	13.0	12.3
7	13.9	13.7	13.5	13.1	12.7	12.3	11.9	11.2	10.5	9.9
8	12.1	11.9	11.7	11.2	10.9	10.5	10.1	9.4	8.7	8.1
9	10.7	10.5	10.3	9.8	9.4	9.1	8.7	8.0	7.4	6.8
10	9.6	9.3	9.1	8.7	8.3	8.0	7.6	6.9	6.3	5.7
11	8.6	8.4	8.2	7.8	7.4	7.0	6.7	6.0	5.4	4.8
12	7.9	7.7	7.5	7.0	6.7	6.3	5.9	5.3	4.7	4.1
13	7.2	7.0	6.8	6.4	6.0	5.6	5.3	4.7	4.1	3.6
14	6.7	6.5	6.3	5.9	5.5	5.1	4.8	4.1	3.6	3.1
15	6.2	6.0	5.8	5.4	5.0	4.6	4.3	3.7	3.1	2.7
16	5.8	5.6	5.4	5.0	4.6	4.2	3.9	3.3	2.8	2.3
17	5.4	5.2	5.0	4.6	4.2	3.9	3.5	3.0	2.5	2.0
18	5.1	4.9	4.7	4.3	3.9	3.6	3.2	2.7	2.2	1.8
19	4.8	4.6	4.4	4.0	3.6	3.3	3.0	2.4	2.0	1.6
20	4.5	4.3	4.1	3.7	3.4	3.0	2.7	2.2	1.7	1.4

Cont.

Period	1% p	1.5% p	2% p	3% p	4% p	5% p	6% p	8% p	10% p	12% p
21	4.3	4.1	3.9	3.5	3.1	2.8	2.5	2.0	1.6	1.2
22	4.1	3.9	3.7	3.3	2.9	2.6	2.3	1.8	1.4	1.1
23	3.9	3.7	3.5	3.1	2.7	2.4	2.1	1.6	1.3	1.0
24	3.7	3.5	3.3	2.9	2.6	2.2	2.0	1.5	1.1	0.8
25	3.5	3.3	3.1	2.7	2.4	2.1	1.8	1.4	1.0	0.7
26	3.4	3.2	3.0	2.6	2.3	2.0	1.7	1.3	0.9	0.7
27	3.2	3.0	2.8	2.5	2.1	1.8	1.6	1.1	0.8	0.6
28	3.1	2.9	2.7	2.3	2.0	1.7	1.5	1.0	0.7	0.5
29	3.0	2.8	2.6	2.2	1.9	1.6	1.4	1.0	0.7	0.5
30	2.9	2.7	2.5	2.1	1.8	1.5	1.3	0.9	0.6	0.4
31	2.8	2.6	2.4	2.0	1.7	1.4	1.2	0.8	0.5	0.4
32	2.7	2.5	2.3	1.9	1.6	1.3	1.1	0.7	0.5	0.3
33	2.6	2.4	2.2	1.8	1.5	1.2	1.0	0.7	0.4	0.3
34	2.5	2.3	2.1	1.7	1.4	1.2	1.0	0.6	0.4	0.3
35	2.4	2.2	2.0	1.7	1.4	1.1	0.9	0.6	0.4	0.2
36	2.3	2.1	1.9	1.6	1.3	1.0	0.8	0.5	0.3	0.2
37	2.2	2.0	1.9	1.5	1.2	1.0	0.8	0.5	0.3	0.2
38	2.2	2.0	1.8	1.4	1.2	0.9	0.7	0.5	0.3	0.2
39	2.1	1.9	1.7	1.4	1.1	0.9	0.7	0.4	0.2	0.1
40	2.0	1.8	1.7	1.3	1.1	0.8	0.6	0.4	0.2	0.1

Index